AUTONOMOUS VEHICLE LIDAR:
A TUTORIAL

I0491002

SREEVATSAN BHASKARAN, KAI ZHOU, ANDREW BAAB, AND RON CALHOUN

Table of Contents

ACKNOWLEDGMENTS

This book grew out of a project with the U.S. Office of Naval Research to provide research experiences for U.S. veterans at Arizona State University. We would like to thank the project managers for the Neptune Program, Rich Carlin and Maria Medeiros, and the many veterans, active military personnel, and professors at our collaborating universities: MIT, Stanford, University of California (Davis), U.S. Naval Academy, Naval Postgraduate School (Monterey), and Purdue – who patiently listened to our presentations on lidar from the beginning. We would also like to thank ASU *Lightworks*, and especially Bill Brandt, Travis Johnson, and Joseph Sanchez who provided mentoring and the administrative lead at ASU for our project. The ASU Pat Tillman Veterans Center, led by Steve Borden, provided outstanding connection between our lidar group and the large veterans community at ASU. We also would like to acknowledge the contributions of Casey Calhoun, who helped with the beam steering laboratory work in the summer of 2019 and Chapter 6 of this book.

CHAPTER 1: MOTIVATION & FUNDAMENTALS

Leading technology companies and a host of startups from all over the world are now competing to solve the challenging problem of autonomy for ground-based vehicles. New startup ventures have emerged from stealth mode and have begun early testing of their autonomous technology in prominent cities across the globe. The development of this cutting-edge technology gained momentum after the 2005 DARPA grand challenge [1]. DARPA's Driverless Car competitions (2004, 2005, and 2007) spurred a series of technological developments and motivated innovation in multiple fields, including artificial intelligence, machine vision, and laser sensors. Familiar characters such as Dave Hall (Velodyne), and Sebastian Thrun (head of Stanford's winning team in 2005 DARPA Challenge) emerged as leaders and went on to found companies. A major innovation was the development of laser sensors for ranging, specifically for autonomous vehicles. These sensors, known as 'lidar' – Light Detection and Ranging, can provide precise range measurements (depth measurements) for cars as they navigate through complex environments. Several startups today focus on providing other sensor solutions like radar, ultrasound, and high-fidelity maps for autonomous vehicles. Today, we see self-driving cars with one or more of these sensors navigating major cities such as Phoenix, Pittsburgh, San Francisco, Shanghai, and Beijing.

1.1 WHY AUTONOMOUS VEHICLES?

Most vehicular accidents are attributed to human error. The fundamental promise of autonomous vehicles is safety, for occupants in cars and for pedestrians and bicyclists potentially sharing the roadways. The ultimate goal is to hand over the entire control of the car to the autonomous driving system. In addition to safety, autonomous driving promises more optimal vehicle usage, potentially reducing parking requirements and traffic congestion. Coordinated and more collaborative driving strategies between large numbers of cars serving urban areas could improve energy consumption and reduce environmental pollution. Transition to fully electric autonomous cars in cities may help reduce emissions of urban air pollution and green-house gases. Autonomous driving systems could relieve the stress of commuting and allow riders greater flexibility and productivity. These advantages motivate the automotive industry to pursue complete autonomy in the next decade, within the safety confines stipulated by the National Highway Traffic Safety Administration (NHTSA).

1.2 AUTONOMOUS VEHICLE - SAFETY LEVELS

Most new cars today have some advanced safety features by default. Advanced Driver Assistance System (ADAS) features are available to aid the driver during parking, lane keeping, blind spot monitoring, or cruise control. While these features can improve safety and reduce driver fatigue, complete autonomy could transform major sectors of our economy. To organize development during the drive towards full autonomy, the NHTSA [2] has proposed a roadmap of "autonomy levels", see Figure 1.1.

The Figure 1.1 clearly lists the preconditions to be met before a given safety level can be assigned to a self-driving car. 'Level 0' is a basic car, where the driver is completely responsible for all the decisions pertaining to the car - no vehicle assistance system is active. 'Level 1' safety is assigned to cars that are still under the purview of the driver - the car's motion is controlled by the driver and has some assistance from vehicle systems (like blind-spot monitoring, lane assistance, or emergency braking). 'Level 2' is commonly referred to as partly automated - the vehicle is responsible for tasks like lane keeping and lane changing (for very specific cases and driving scenarios) and the driver is required to continuously monitor these systems. 'Level 3' systems handle scenarios as in 'Level 2' - they are also capable of alerting the driver to take over when facing a situation that is beyond their capabilities (the driver is required to constantly monitor these safety systems and take-over control when alerted). 'Level 4' entails autonomy where human

2

intervention is required when encountering a new situation. Finally, 'Level 5' is a driverless system that does not need human intervention at any time.

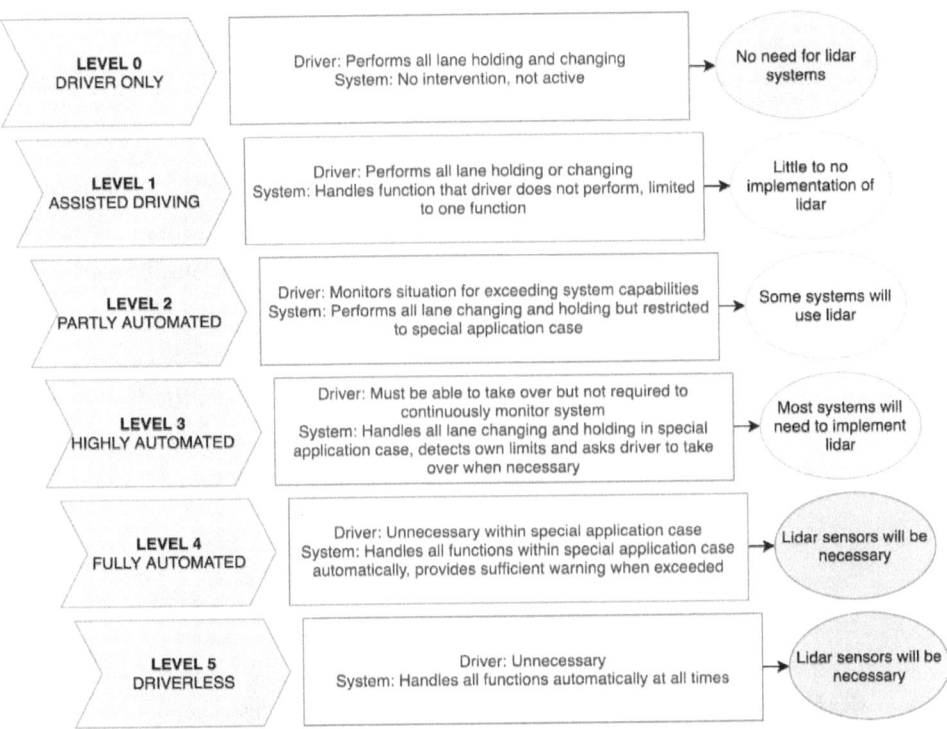

Figure 1.1 Levels of autonomy (left) and our expectations for lidar (right).

1.3 SENSORS FOR AUTONOMY - PERCEPTION SYSTEM

Advanced driver assistance systems in self-driving cars use lidars, cameras, and radars to sense the environment around the car. The information from the sensors can be used to initiate a sequence of actions by the autonomous system to control the car's trajectory and behavior. For example, an emergency braking system monitors the space in front of the car using cameras and radars to identify obstacles and apply brakes. Cameras provide images with very fine detail but the images lack depth information. Radars can extract both depth and velocity information of the target of interest. They can "see" longer distances, 300 m, for example, which gives the autonomous system enough time to respond to obstacles in its path. However, locating an obstacle within the field of view with high angular precision is challenging for a radar system. Lidar is an optical sensor which uses coherent light to measure depth information with very high angular resolution. Also, consecutive range estimates from the lidar sensor (along a given line of sight) can be used to extract target velocity.

A driverless car has three major sub-systems [3] as shown in Figure 1.2, A) <u>Perception</u> – the sensor suite, e.g., lidar, radar, camera, or ultrasound sensor, B) <u>Intelligent Agent</u> – The "brain" of the autonomous system. It uses the data from the perception sub-system as input to detect and track obstacles around the vehicle. Based on these detection results it commands the car's actuators. C) <u>Control System</u> receives the instruction from the intelligent agent and executes the corresponding actions on the automobile, e.g., controlling engine throttle, steering the wheel, signaling, or braking.

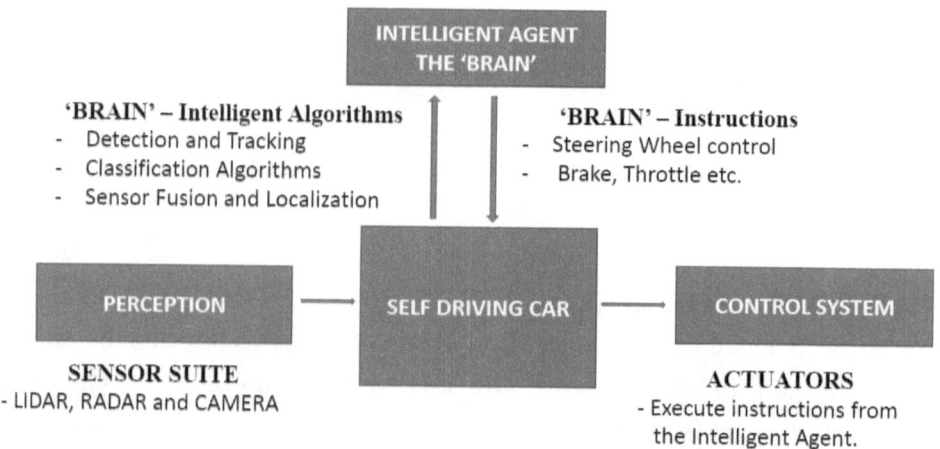

Figure 1.2 Block diagram: Self-driving car sub-systems

The perception sub-system is critical to the overall performance of the self-driving car. Using the on-board sensors, it must map the environment (with both static and dynamic objects) around the vehicle with high precision and accuracy. However, variations in the weather is a challenge for accurate range sensing.

This sub-system provides a high-fidelity depth map which is necessary for localization tasks. Localization is the process of estimating the position of the self-driving car in the global map using known landmarks as markers. This improves tracking dynamic objects around the car.

To enable safe navigation for a self-driving car its sensor should produce a depth map with very high resolution and be capable of locating targets over a specified range. A normal Global Positioning Sensor (GPS) provides location estimates with very poor accuracy, typically to the order of few tens of meters. On the other hand, a lidar sensor reduces the local range error to the order of few centimeters. This, when properly combined with coarser (but more global) navigational mapping, can essentially provide a form of 'digital rails'. Further, since both dynamic and static objects are scanned, lidar provides a basis for timely response to rapidly developing transient challenges (e.g., bicyclist unexpectedly crossing against signals). Most leading AV teams utilize a suite of sensors, fusing data from multiple sensors to obtain features such as range, color or velocity with high accuracy and precision. A data fusion approach that uses complementary sensors, attempts to addresses limitations inherent in any single sensing modality. This provides redundancy and consensus which sharpen detection certainty and tracking in real time.

For lidar, rain and fog can be quite limiting [4]. The transmitted optical signal is attenuated by its interaction with water molecules, thus degrading lidar performance. In such a scenario, multiple sensors (camera, radar, infra-red sensors, ultrasonic sensors) must be leveraged to navigate the complex terrain safely. Also, the size and cost of a lidar sensor is a major bottleneck that has driven engineering researchers to explore approaches to mitigate these challenges. Large-scale production will bring down lidar prices for automotive applications.

Typically, lidar sensors are categorized based on their range of operation:

a. **Short range lidars** → Measure ranges up to 20 - 40m
b. **Medium range lidars** → Measure ranges up to 100 - 150m
c. **Long range lidars** → Measure ranges up to 200 - 300m

Short range lidars can be used for parking assist applications in cars. It is also extensively used by robotics researchers, enthusiasts and mapping experts for indoor applications. Medium range lidars are necessary for monitoring road intersections and while exiting/merging onto a highway. Other applications include mapping streets in urban areas where medium range visibility is required to identify pedestrians and bicyclists on the road. Long range lidars are necessary when travelling at very high speeds. For example, while travelling at 75 mph on a highway/expressway lane, identifying obstacles at 200 – 300m range is critical. This gives the autonomous system about 6-8 seconds to respond and decide on the car's behavior over the next few critical seconds. A combination of short, medium and long range lidars is necessary for a self-driving car to navigate safely.

1.4 LIDAR SENSOR MODALITIES

Lidar is a vital sensor necessary to achieve any level of autonomy in vehicles today. It is the "eyes" for an autonomous system. Almost every self-driving car that we see on road today has multiple lidar sensors on board. These sensors produce point cloud data (3D depth information) for the vehicle to process in real time and identify potential obstacles in its path. In this section, the focus will be on the fundamental physics behind the workings of a lidar sensor.

1.4.1 LIDAR SENSOR FUNDAMENTALS

A lidar sensor has the following basic components.

a. **Laser Source** → e.g. Semiconductor, Diode, Fiber Laser.

b. **Laser Amplifier** → e.g. Master Oscillator Power Amplifier, Q-switched amplifier, Erbium Doped Fiber Amplifier.

c. **Transmitter and Receiver Optics** → e.g. Collimating Lens, Plano Convex Lens as receiver.

d. **Scanner** → e.g. Rotating mounts, MEMS mirror, Optical Phase Arrays, Polygonal mirrors, Galvo-mirrors etc.

e. **Optical Detector** → e.g. Photodiode, Avalanche Photodiodes (Linear Mode, Geiger Mode), Focal Plane Arrays.

f. **RF-Amplifier** → e.g. Resistive and Capacitive Transimpedance Amplifier (R-TIA and C-TIA)

g. **Timing Solution** → e.g. Time to Digital Converter, Analog to Digital Converter, Phase detector.

Each of the above components will be explored in detail in subsequent chapters.

There are different types of lidar sensors available today:

- **A direct detection based Time of Flight (ToF)** lidar sensor measures depth information.
- **Frequency Modulated Continuous Wave (FMCW)** lidar sensor estimates both target distance and velocity simultaneously.
- **Amplitude Modulated Continuous Wave (AMCW)** lidar sensor estimates target depth information.

Every lidar sensor manufacturer lists the lidar system critical parameters and its calibrated operating modes in detail. These parameters can be used to compare the performance of different sensors and choose the right sensor modality for a given application. A few critical lidar sensor parameters are discussed below [5]:

a) **Eye Safety**: Lidar sensors shoot laser beams as they scan the free space. Lasers with certain power levels could be harmful to the naked eye. Eye safety standards categorize lidar sensors based on their output optical power level. For automotive applications the lidar sensor must either be classified as a Class 1 or Class 1M laser safety level.

b) **Field of View (FOV):** For automotive applications sensing the environment around the vehicle is critical. In order to map the entire region around the car, a lidar sensor with 360° horizontal field of view (HFOV) is required. One way to achieve this is to mount the sensor on top of the car. A classic example is the Velodyne HDL 64E sensor. It has a vertical field of view (VFOV) of 26.9 ° and a 360° horizontal field of view. Thus, a single sensor is able to map the environment around the car and provide real time point cloud depth data. Another common approach is to use multiple lidar sensors with each sensor sweeping through a sector. For example, if the HFOV is set at 120°, three independent lidar sensors would complete the entire 360° requirement.

c) **Angular resolution:** The angular resolution of the sensor defines the number of points the lidar sensor can map within its FOV. For e.g. the Velodyne HDL 64E sensor has the horizontal angular resolution at 0.08° and a vertical angular resolution at 0.4°. To be able to track features like the edge of a curb or the foot of a bicyclist we need a very fine angular resolution both in the horizontal and vertical axes.

d) **Number of Points:** The FOV and angular resolution will directly tell us the number of points the sensor is capable of mapping within its FOV. Most lidar sensors can scan up to a million points.

e) **Frame rate:** The frame rate for a lidar sensor is very similar to that of a camera. For lidar, it is the number of complete scans over the entire FOV, that the sensor can complete in a second. For example, the Velodyne HDL 64E, a spinning lidar rotates at about 5-20 Hz. It can complete 5 – 20 scans a second. This is a user specified parameter and any number within this range can be specified. Given the number of points for a sensor and its frame rate, the rate at which the point cloud data is produced by the lidar sensor can be computed. This will help to choose a data transfer protocol for basic data storage considerations on a car.

f) **Maximum Range:** A very common specification that is found on a lidar sensor data sheet is the maximum range that the sensor can measure. The maximum range listed is based on tests performed with known targets. The maximum range is a function of the target reflectance typically specified in the sensor datasheet. A reflector is characterized as a poor reflector if its reflectivity is 10% or less, and as a strong reflector if its reflectivity is 90% or more.

g) **Pulse Repetition Rate/Pulse Repetition Frequency (PRR/PRF):** The PRR/PRF is a laser source parameter. It specifies the number of pulses (for a pulsed lidar) that the laser diode can produce in a second. The number of points within the FOV sets the PRF requirement for the lidar sensor. The PRF of the system inherently sets a physical limit on the maximum range that can be probed by the sensor.

h) **Peak Power/Average Power:** Peak power is the amount of energy packed into a single pulse of known pulse width. The average power is the energy spread over a duty cycle defined by the PRF. The laser peak power and average power are design parameters for a lidar system that have to be within the eye-safety norms. The pulse width and the PRF of the laser source are parameters that can be used to estimate the peak power and average power given the energy per pulse.

i) **Wavelength**: The wavelength of operation is another important lidar sensor parameter. The laser beam (pulsed or continuous wave) is attenuated as it propagates through the atmosphere. The photons are absorbed by water molecules in the atmosphere and thus reduce the strength of the return signal. Most common wavelengths are 905 nm, 1550 nm and 850 nm. Velodyne uses a laser diode in the 905nm range. Luminar on the other hand uses a 1550 nm laser source.

j) **Range Resolution**: This is the most important specification for a lidar sensor. It is a measure of the smallest change in range/depth

9

that is perceivable by the lidar system. It is a strong function of the return signal strength, the sensitivity of the photodetector and the resolution of the timing device. Similarly, accuracy of the depth measurement by a sensor (a measure of how close the estimate is to the true depth value) and the precision (a measure of the depth estimate uncertainty) are important parameters to compare different lidar sensing modalities for autonomous applications.

In the next section, we discuss different lidar sensor modalities [6-9] and their underlying physics of operation.

1.4.2 DIRECT DETECTION PULSED LIDAR SENSOR

This is the most common type of lidar sensor available today (Figure 1.3). In a direct detection system, the laser source produces short bursts of optical energy. Typically, a very narrow pulse (pulse-width to the order of a few picoseconds or nanoseconds) is generated by the laser source at a user specified PRF. The optical pulse is then amplified by the laser amplifier. These pulses are then directed into the atmosphere with the help of a scanning device. Each pulse is directed to a specific location within the sensors field of view as it scans over the region of interest. A portion of the backscattered pulse energy that reaches the receiver lens is then collected, amplified and converted to an equivalent electrical voltage signal.

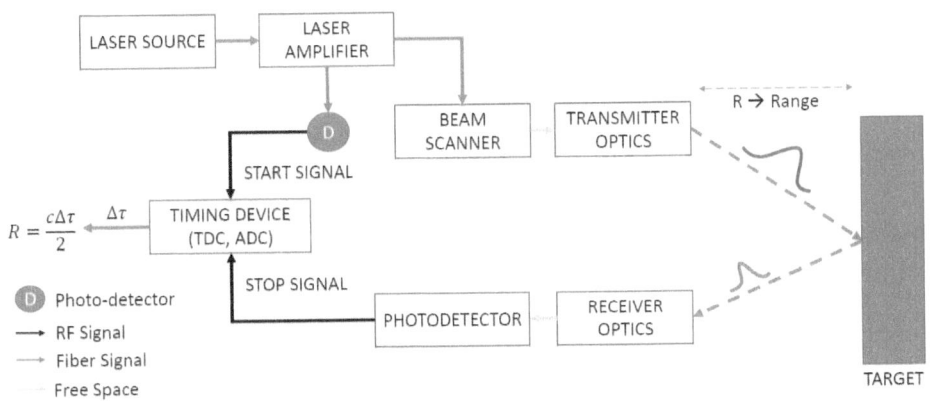

Figure 1.3 Direct detection (TOF) lidar block diagram

For pulsed lidar systems a very fast photodetector i.e., a large bandwidth detector is required to capture the return pulse and its rise time accurately. A very precise clock, typically a Time to Digital Converter (TDC), measures the time between the pulse transmission event and pulse arrival event based on a simple leading-edge detection technique [8].

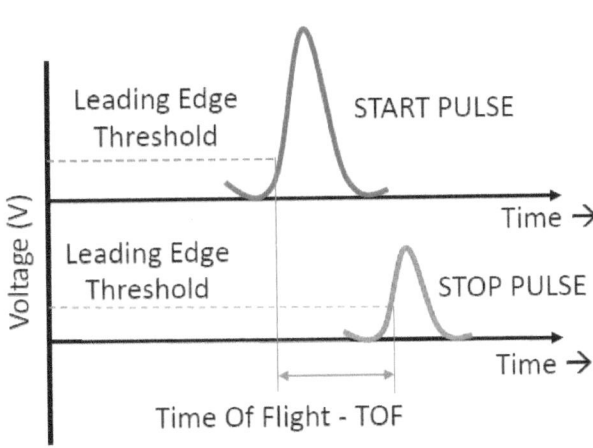

Figure 1.4 TOF Lidar start-stop pulse with leading edge detection threshold

The measured time difference is converted to range information with the known speed of light (2.99 x 10^8 m/s) using Equation 1.1. (Figure 1.4 – START and STOP Pulse)

$$R = \frac{c\Delta T}{2}$$ (1.1)

where 'R' is the range, 'c' is the speed of light, and 'ΔT' is the time of flight (TOF).

A pulsed lidar sensor is also commonly referred to as a time of flight (TOF) device. The smallest change in range that can be measured by the TOF device is dependent on the smallest time difference the clock can capture between the transmitted and the received pulse. Thus, the range resolution 'ΔR' of a TOF device is dependent on the clock's timing resolution (Equation 1.2). A nano-second pulse, large transmitted pulse peak power and a high-resolution clock is required for accurate and precise TOF measurements. For example, to achieve a range resolution of approximately $\leq 2\ cm$, the timing resolution needs to be $< 0.13\ ns$ [9], given by Equation 1.2.

$$\Delta R = \frac{c\Delta T}{2}$$ (1.2)

For a pulsed lidar, the repetition rate sets the limit on the maximum range that can be measured. The pulse repletion rate for example at 1MHz, sets the maximum range at 150 meters (corresponding to a $1\mu s$ round trip time of flight, PRF $= 1$ MHz $\Rightarrow 1\mu s$ between consecutive pulses). This limit is to avoid the interaction between the returns from a pulse and the next subsequent pulse transmitted from the laser source.

1.4.3 Frequency Modulated Continuous Wave (FMCW) Lidar

FMCW lidar (Figure 1.5) uses a tunable laser source [10-18], i.e., the laser source transmits a signal that has a range of frequencies that varies over time. The laser frequency is linearly chirped as shown in

Figure 1.6. The linear chirp can be produced by driving the laser diode with a waveform generator. The time duration of the linear frequency ramp is given by 'T_s'. Every single range measurement needs to be completed within this sample time duration. The return signal is shifted in frequency by an amount that is proportional to the target range. These lidars use the coherent heterodyne detection technique to track the change in frequency between the transmitted and the received signal. A reference local oscillator signal, from the laser source, is used to coherently mix with the return signal using a square-law photodetector. The output from the photodetector is proportional to both the incident optical power and the square of the magnitude of its electric field. The oscillating frequency of the photodetector output (AC component) is the beat frequency to be estimated (Figure 1.5).

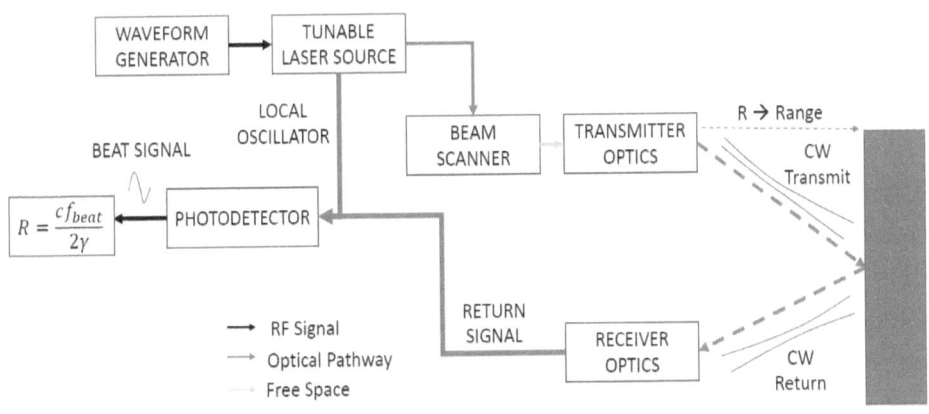

Figure 1.5 FMCW Lidar Block Diagram

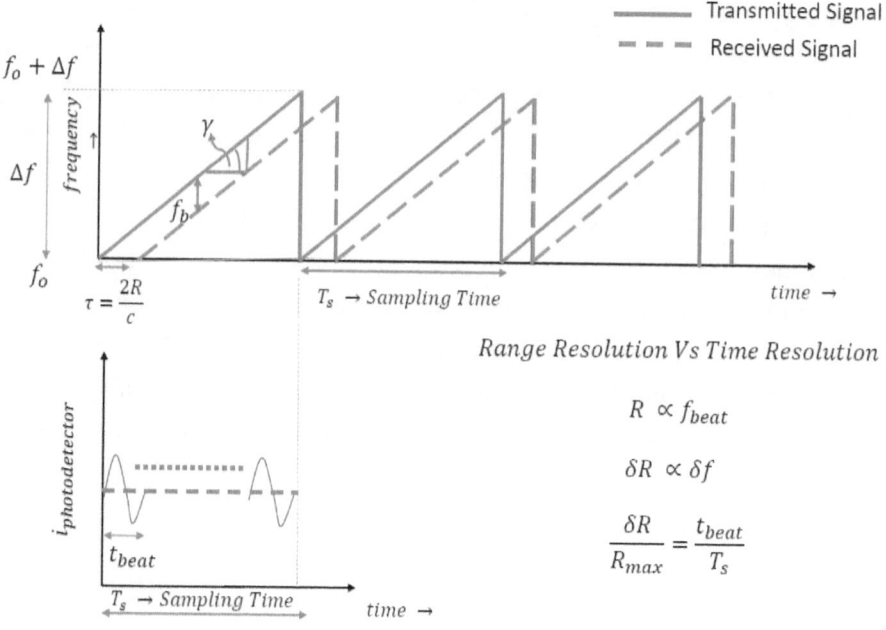

Figure 1.6 FMCW lidar laser source linear chirp (saw-tooth) and photodetector output as a function of time [11].

The measured beat signal frequency is used to estimate the distance to the target (Equations 1.3a&b). Heterodyning steps down the bandwidth of the return signal to the baseband range i.e., in the kHz to MHz range. Thus, the optical signal can be converted to a current or a voltage signal using a lower bandwidth detector (slower detector). The return signal is contaminated by various noise sources, one of which is the photodetector itself. A lower bandwidth detector contributes to a lower magnitude of random noise to the return signal. Also, the detector output signal amplitude is proportional to the product of the amplitude of the return signal and the local oscillator [11]. This implicit amplification of the return signal improves the detection of weak signals. Unlike a pulsed lidar system, a very high peak power transmitted pulse is not necessary as the FMCW approach focuses on tracking the beat frequency and not the amplitude of the return signal.

14

$$R = \frac{f_{beat}\, C}{2\gamma} \qquad\qquad (1.3a)$$

$$R = \frac{f_{beat}\, C\, T_s}{2\Delta f} \qquad\qquad (1.3b)$$

where 'R' is range, 'f_{beat}' is the beat frequency, 'C' is the speed of light and 'γ' is the slope of the frequency modulation curve($\gamma = \frac{\Delta f}{T_s}$).

The beat signal frequency is measured by calculating the period of the oscillations as shown in the Figure 1.6 [11]. The maximum range that can be measured is dependent on the largest beat frequency that can be measured. This requires the FMCW lidar system to measure very small beat signal time periods (Figure 1.6) of the current signal (t_{beat}). Similarly, for range resolution, the smallest frequency that can be measured sets the limit. The smallest frequency that can be measured corresponds to the largest beat period that can be measured, i.e., T_s.

For example, let the measurement duration (T_s) be $1\mu s$. If we want to measure a maximum range (R_{max}) of 1 meter with a precision of (δR) $10\mu m$, then the clock timing resolution is $1\,ps$ (Equation 1.4 - normalized range resolution and time resolution). For the same scenario to achieve a precision of $10\ \mu m$ a pulsed laser system would require a clock resolution of approximately $6.7\ fs$. Thus, the requirements on the clock and its timing resolution is relaxed for a higher range precision and accuracy promised by FMCW lidars.

$$\frac{\delta R}{R_{max}} = \frac{t_{beat}}{T_s} \qquad\qquad (1.4)$$

Ideally the returns from the target should produce a current signal with a single tone (beat frequency). However, laser diodes have a non-linear response to the frequency modulation. The frequency chirp as a function of time is not linear as shown in Figure 1.6. This produces an additional phase noise to the transmitted signal. Also, the coherence length of the laser diode limits the performance of the FMCW lidar. The coherence length of laser diode is a measure of its coherence time (Equation 1.5) [12].

$$L_{coherence} = c\tau_{coherence} = \frac{c}{\pi\Delta v} \qquad (1.5)$$

where Δv is the optical bandwidth (FWHM – Full width at half maximum), $\tau_{coherence}$ is the coherence time of the laser.

A low value implies that the phase information decorrelates quickly due to the phase noise produced within laser diode. The process of coherent heterodyning requires the return signal and the transmitted signal to have a strong phase correlation for effective constructive interference at the photodetector. With the phase noise from the laser diode creeping into the transmitted signal, the returns are no longer a single tone beat frequency. The return has additional modes that deteriorates the accuracy and restricts the maximum range that can be measured by a FMCW lidar.

Another configuration of an FMCW lidar uses a triangular waveform [10] (Figure 1.7) instead of a saw tooth waveform for the laser frequency modulation as shown in Figure 1.6. The sawtooth waveform is used only to measure range information for stationary objects. The triangular waveform has the added capability to measure velocity in addition to range information of targets within the field of view. When tracking a moving target, two beat frequencies embedded in the return signal are estimated. The velocity of the moving target is given by the difference between the two beat frequencies (Equation 1.6), and the range to the target is given by the average of the two beat frequencies (Equation 1.7).

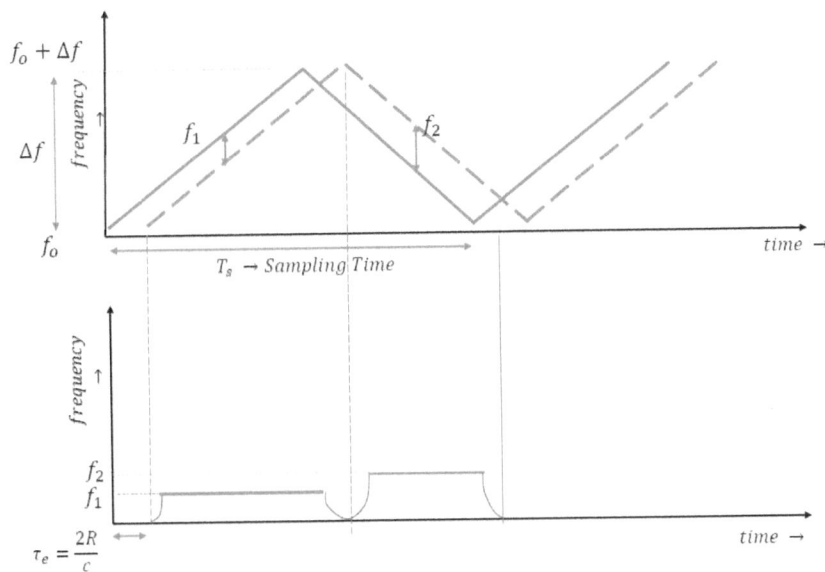

Figure 1.7 Triangular waveform for laser source frequency modulation [10].

$$f_{doppler} = \frac{f_2 - f_1}{2} \qquad (1.6)$$

$$f_{range} = \frac{f1 + f2}{2} \qquad (1.7)$$

$$R = \frac{(C\ T_s\ f_{range})}{4\Delta f} \qquad (1.8)$$

$$V_r = \frac{2f_{doppler}}{\lambda} \qquad (1.9)$$

where V_r is the radial velocity, λ is the wavelength of the CW laser, τ_e (Figure 1.6) is the time duration during which the transmitted signal travels to the target and back.

If the range and the velocity of the target is a constant during the measurement time T_S, the range estimate and the velocity (component along the direction of the beam) are given by Equations 1.8 and Equations 1.9. Thus, FMCW lidars promise the simultaneous measurement of velocity and range. Optical heterodyning provides FMCW lidars the capability to be resistant to stray light and background light. The reference and the return signals should have a decent phase correlation to ensure efficient mixing. Incoherent background light will no longer contribute to the photodetector output.

1.4.4 Amplitude Modulated Continuous Wave (AMCW) Lidar

AMCW lidars [19-25] (Figure 1.8), like FMCW lidars, are typically used for short range applications, i.e., to measure distances between a few centimeters to a few tens of meters. An AMCW lidar measures the phase difference between the transmitted signal and the return signal (Figure 1.9). The phase shift in the return signal with respect to the transmitted signal is proportional to the range of the target. It is given by Equations 1.10, 1.11 and 1.12.

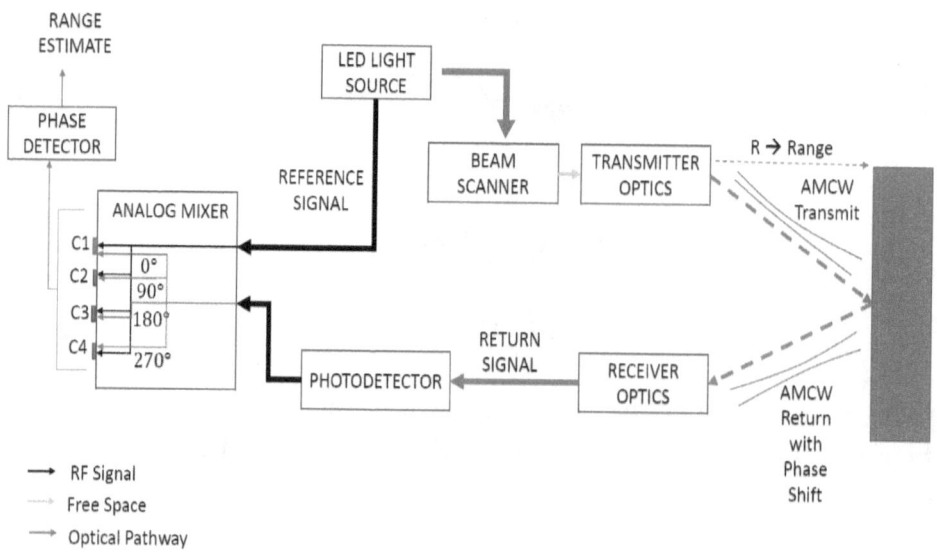

Figure 1.8 AMCW Lidar Block Diagram.

$$\phi_{diff} = \frac{2\pi}{\lambda_{mod}} \Delta x \qquad (1.10)$$

$$\phi_{diff} = \left(\frac{2\pi}{\lambda_{mod}}\right) 2R \qquad (1.11)$$

$$\phi_{diff} = \left(\frac{4\pi}{\lambda_{mod}}\right) R \qquad (1.12)$$

where ϕ_{diff} is the phase difference, Δx is the path difference – round trip distance, R is the range to the target, λ_{mod} is the modulation wavelength related to the modulation frequency f_{mod} as given by Equation 1.13,

$$f_{mod} = \frac{c}{\lambda_{mod}} \qquad (1.13)$$

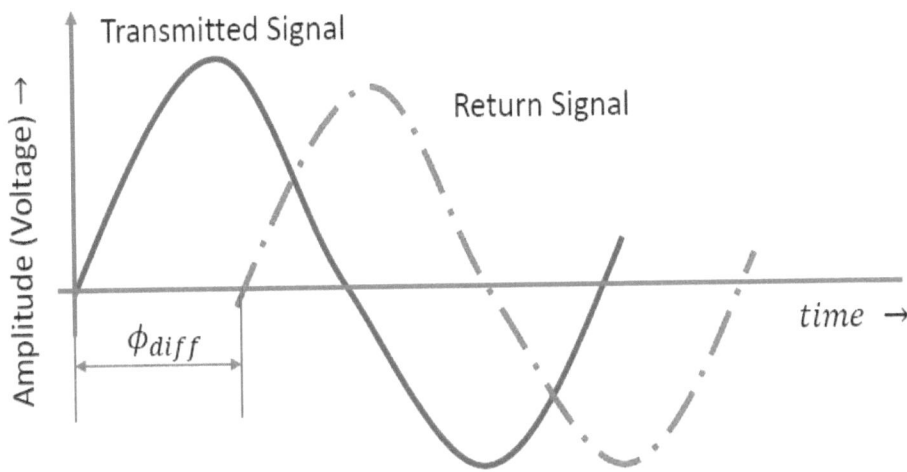

Figure 1.9 AMCW lidar transmitted and return signal phase difference [24]

For example, if the modulation frequency is 10 MHz, then the modulation wavelength is 30 m. The range resolution of an AMCW lidar is a function of the phase resolution of the system. A phase detector is used to measure the phase difference between the transmitted and the received signal. From the Equation below (Equation 1.14), for a given phase resolution, the range resolution can be lowered by using a lower modulation wavelength (larger modulation frequency).

$$\Delta R = \Delta \phi_{\text{diff}} \left(\frac{\lambda_{mod}}{4\pi} \right) \qquad (1.14)$$

For an AMCW lidar, the modulation frequency limits the maximum range that can be measured i.e., it sets the maximum ambiguous range. The maximum phase difference that can be measured is 2π. Thus the maximum range that can be measured is given by,

$$R_{max} = \frac{\lambda_{mod}}{2} \qquad (1.15)$$

Using the Equation above, with $\lambda_{mod} = 30m$, the maximum un-ambiguous range that can be measured is $15\ m$. For practical applications, the maximum range will be less than this maximum limit owing to beam attenuation and target reflection characteristics. For a lower modulation wavelength, the range resolution will improve at the expense of the maximum range that can be measured using the AMCW lidar system. This is a very important design trade-off for AMCW systems. One way to achieve better range resolution and simultaneously ensure a larger maximum range is to use two frequencies in the amplitude modulated beam. The lower frequency component will improve the maximum range limit, and the higher frequency component will ensure better range resolution. An approach to design such a system is described in reference [24].

A single frequency AMCW lidar transmits an amplitude modulated signal into the atmosphere and generates a reference signal (analog). The output of the photodetector is now mixed with the reference signal shifted in phase, typically at $0°, 90°, 180°\ and\ 270°$. The mixed signal is sampled over a specified integration time and accumulated to improve the return signal SNR (Signal to Noise Ratio). The samples from the four channels (Figure 1.8) are used to compute the phase difference given by the Equation 1.16 [23].

$$\phi = \arctan \left(\frac{C3 - C4}{C1 - C2} \right) \qquad (1.16)$$

where C1, C2, C3 and C4 are the signal samples from the four channels. The phase difference between the transmitted and the received

20

signals can be estimated using the Equation above. Ideally the return signal would be a clean sinusoid with a phase shift with respect to the transmitted beam. However, for practical applications, the return signal might have additional harmonics in it that arises due to multipath interference effects. Using suitable filters in the return signal chain, frequencies outside the required band can be filtered out. Also, the return signal amplitude (the signal to noise ratio) can be improved by increasing the integration time to estimate C1, C2, C3 and C4. However, this may introduce motion blur effects due to dynamic objects in the scene.

1.5 CONCLUSION

There are different types of lidars available in the market today that are suitable for different applications. Understanding the sensor specifications will help in evaluating different sensor modalities. These system/sub-system level characteristics of the sensor are critical to designing our own lidar sensor as well.

In the next chapter, the focus will be on the design considerations for direct detection lidar systems. A systems level approach will be adopted to address how one selects the right components to realize a high-level sensor design.

CHAPTER 2: TOF SYSTEM DESIGN - LIGHT PATHWAY

2.1 EXAMPLE DESIGN CONCEPT

In the last chapter, the basic physics of different lidar sensor modalities were presented. In this chapter, we focus on the design and concepts for a TOF lidar system, using a systems engineering approach [26]. Our goal is to construct a TOF lidar sensor using off the shelf components available in the market today. In the course of our design thinking, we will consider and illustrate governing physics, optics, and engineering guidelines which are most relevant for autonomous vehicle lidar. (We expect that different viewpoints or intended market niches or opportunities would lead to different designs, while still being governed by the same physical and engineering principles.)

We begin with answering a fundamental question: *What is the role of the lidar that we would like to design?* A lidar could be built, from the ground up, as a companion for and complement to camera-based sensors. Noting the success of essentially "camera-like" sensors (our eyes!) which provide input to human drivers, one might argue that cameras and image processing should be the "work-horse" of the sensor system. Nevertheless, "disambiguating" - difficult to discern objects or conditions in the field of view, perhaps at long range, or in poor lighting, could be the design purpose for a lidar. One might believe, for example, that a large fraction of vehicle accidents is associated with poorly classified objects with unclear trajectories. Therefore, the role of the lidar could be to provide a more precise "second-look" at key sub-regions of the field of view. This will be our example concept, from which we will illustrate design principles.

Developing this concept further, we seek to assess whether this role realistically capitalizes on the best of current technology and fills a technical need for autonomous vehicles. Sensing with precision at longer ranges (e.g., 200 m) should be well-suited to inherent strengths of lidar due to low beam divergence. Therefore, a lidar should provide good angular resolution (presuming, however, that beam steering is sufficiently accurate) and concentrated delivery of energy to distant objects. Taking more inspiration from the example of human drivers, we notice that key regions of the field of view attract disproportionate levels of attention. That is, objects of importance with trajectories or positions that could intersect with a driver's vehicle are focused on in more detail. For example, the corner of a curb, around which a driver wishes to center a turn in a parking garage, attracts a more deliberate gaze.

A lidar designed specifically for a "more deliberate gaze" would need to be able to move quickly and steer a laser beam flexibly, with an ability to arbitrarily "paint" boundaries of distant objects. We note at this point in the design that MEMs mirrors would seem a promising candidate for beam steering, due to very low inertia, rapid movement, and ability to maintain the integrity of the laser beam. Secondly, a study of laser sources at the time of writing reveals current availability of high peak power, small-footprint, energy efficient fiber lasers at IR frequencies.

Reviewing our concept for a lidar, we expect that the AI will be the center-point of autonomy, controlling sensors and assimilating their data streams. Specifically, for the lidar, the AI would mark regions from camera images for enhanced probing and accept 3D range point cloud data (concentrated preferentially on these key objects) as a basis for refinement of its classifications. We need, therefore, precision beam steering and efficient delivery of high peak powers to longer ranges. Although there are many IR frequencies that could be considered, a survey of the literature shows that 1550 nm would seem favorable for eye safety reasons. Furthermore, at wavelength of 1550 nm, both laser sources and detectors are at a high level of technological readiness, due primarily to developments in the telecom industry and in military rangefinders.

We begin with the power in the laser beam, proceeding from the component that produces the optical power, to how it is estimated along its optical pathway, and finally to how much we can expect to return to our sensor.

2.2 RADIOMETRY- INTRODUCTION TO POWER BUDGET

As stated previously, the first step was to identify a niche role for the lidar sensor. Having fixed the functionality of the lidar sensor, i.e., to disambiguate regions of interest within the sensors field of view, the next step in the lidar sensor system design is to ask, how far do we want to see (maximum range) with our sensor? Most lidar sensors available today market themselves as sensors that can range up to 120 meters (for e.g. Velodyne HDL-64E) [57]. It is required that we evaluate each sensor carefully, identifying and clearly understanding the conditions under which the marketed specifications can be met efficiently.

The goal here is to design a long-range sensor that can see up to 150 m. This gives the autonomous vehicle enough time to decide its behavior given a potential threat in its path has been identified at this range. Thus, it is critical that the autonomous system is given as much time as possible, i.e., the sensor should be ideally capable of seeing far ahead of the vehicle to respond timely to changing dynamics of the environment.

To begin, we need a power budget that can help us evaluate the amount of optical power to be transmitted and the amount of optical power returned from the target. A *MATLAB*™ model can be built to perform trade off studies that will lead to the right choice of laser source, transmitter receiver optics, and most importantly the receiver sensor. The fundamental equation driving this model is the Lidar Equation. In the following sections, each individual subsystem of the lidar sensor will be explored in detail, leading up to the Lidar Equation itself.

2.3 PULSED LASER SOURCE

The maximum range/depth measurement specification for the lidar prototype was set at 150 m. The laser source is required to output enough photons to successfully get enough returns back to the receiver for an accurate measurement. The laser source parameters that need to be considered are its wavelength, peak power, average power, energy of a pulse, pulse width and the pulse repetition rate.

a) **Wavelength** – Lidar sensors currently available in the market today predominantly use wavelength in the 905 – 914 nm range. A handful of new emerging technologies focus on using 1550 nm laser sources. The choice of wavelength is very critical. It is fundamentally driven by the absorption spectrum (Figure 2.1) [27] of the atmosphere.

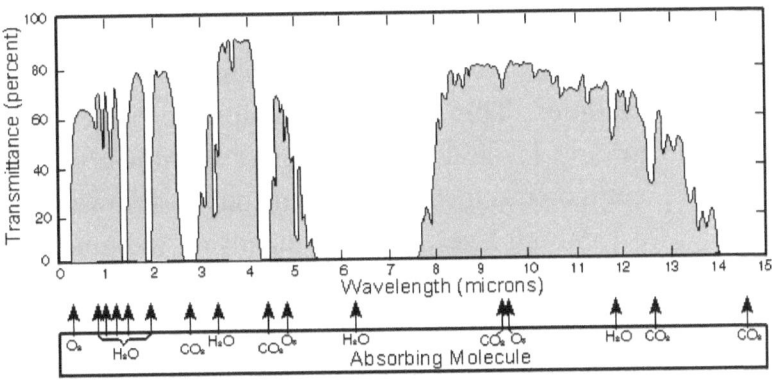

Figure 2.1 Atmospheric Transmission as a function of wavelength. Range assumed in the plot is 1.8 Km (1 Nautical Mile)

The wavelengths that have lower absorption in the atmosphere due to water droplets/moisture are preferred to transmit laser photons effectively over longer ranges. Delivering more photons per pulse, might seem like the right approach to be able to range longer distances. However, eye-safety regulations stipulate an upper limit to the amount of optical power that can be pumped out by the laser source for a given wavelength of operation [53]. A 1550 nm laser pulse, by this eye-safety norm, has a higher upper limit when compared to the 905 nm laser pulse [29]. The human eye has the capability to absorb optical energy at 1550 nm and avoid focusing it onto the retina. However, 905 nm wavelength photons are focused by our eye lens and can cause permanent damage to our eyes if exposure levels are not within the acceptable levels. Also,

1550 nm laser source can take advantage of the mature fiber communication techniques that are perfected for this specific wavelength of operation. At 1550 nm, the optical power loss in a fiber is very low, allowing effective communication over long ranges and very fast internet that we use today. Fiber based laser amplification (erbium-yttrium doped fiber amplifier) at 1550 nm is also a technology that is available off-the shelf today [26]. A simple silicon photodetector can be used to detect 905 nm photons and thus easily integrated onto CMOS chips. These chips can be manufactured in large quantities using existing manufacturing techniques and facilities. This makes it possible to achieve cheap volume pricing levels that the automotive industry seeks to get the autonomous technology to the public at lower costs. However, a 1550 nm laser source demands an Indium-Gallium-Arsenic detector that has very high responsivity at 1550 nm [26]. When compared to the 905 nm technology, this is more expensive. So, it demands research and development to help innovate new techniques. Lidar sensors from Luminar, Strobe and Blackmore have come out with their sensor solutions that uses a 1550 nm laser source [30-32]. The next generation lidar sensors will eventually need to tap into this space to be able to develop long-range lidar sensors. Example calculations in this book explore a 1550nm laser source to meet our long-range specification.

b) **Laser Power** – The amount of optical power output from a laser source is characterized by its peak power, average power and the pulse energy. Peak power is the ratio of pulse energy to the width of the pulse (Full Width at Half Maximum - FWHM). Thus, with the pulse energy fixed, using a smaller pulse-width laser source produces a higher peak power pulse. For ranging sensors that depend on the strength of the return, i.e., direct detection-based sensors, the name of the game is to deliver maximum instantaneous optical power to the target.

Thus, using a shorter pulse-width, a higher peak power can be delivered, and a stronger return can be expected from the target. Most commercially available laser sources have capabilities to produce nanosecond or picosecond pulses. For our lidar prototype, a laser source with variable pulse-width (5ns – 20ns) is used. Average power is the product of the energy of the pulse and the pulse repetition rate (PRR). A laser source should produce pulses at high frequencies to meet the requirements on the number of points that the sensor needs to map within its field of view.

Different types of laser sources are available today. A driver circuit drives a current signal (at a specified peak amplitude and duty cycle) through a diode laser to produce a pulse (with the required peak power, pulse width and PRR). The optical pulse produced is amplified using a laser amplifier to increase the optical power to the requisite level (within eye-safety limits). For the lidar prototype, a 1550 nm distributed feedback (DFB) laser source is used to produce pulses at 20 mW peak power. An erbium-yttrium doped fiber amplifier is used to amplify the output pulse peak power to approximately 381 W at a user specified pulse width of 5ns and pulse repetition rate of 750 MHz A narrow linewidth laser source is required to ensure strong coherence of the output laser beam. The linewidth is typically a measure of the width of the spectrum about its center frequency which in this case corresponds to 1550 nm. The laser source used has a narrow linewidth of ± 10nm.

2.4 TRANSMITTER OPTICS

With a 1550 nm laser source and a fiber-based amplifier, the lidar system developed classifies as a fiber-optics based TOF lidar system. That is, the optical pulse generated by the laser source is input into the amplifier (fiber based) and finally the amplified optical pulse is transmitted into a collimating lens that focuses the optical beam at infinity. The transmitter optics attempts to capture the light from the fiber optic cable efficiently and transmit the optical beam with very low divergence. The fiber cables have typically very low insertion losses, that are quantified and listed on the fiber optic cable datasheet.

The output from the laser source, a laser diode or an optical fiber cable, diverges very quickly as it exits the source. Fiber cables typically have an angular cut at their tips (FC/APC type) [33] so that the beam diverges as it emerges out without harming any user while operating with a high-power laser source. The collimating lens collects this diverging beam and produces a parallel tight optical beam, with minimum divergence, that is transmitted into free space. A standard choice for laser collimation is a plano-convex lens. The figure below shows a simple setup with a laser diode placed at the focal point of the plano-convex lens. Note that the diode can be replaced by a fiber cable tip as well.

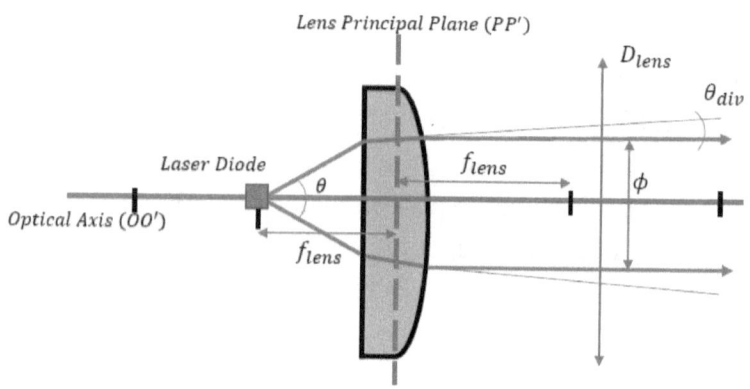

Figure 2.2 Plano-Convex lens parameters.

In the figure above, a diode is placed at the focal point of the lens (f_{lens}). For collimation, the diode is placed behind the planar surface of the lens. The divergence (θ_{div}) of the output beam is a standard metric that is listed on the product data sheet. The laser output from the diode may have a different spread/divergence along the horizontal and the vertical axes, producing an elliptical beam as output. The numerical aperture of the diode is a measure of this divergence (Equation 2.1) [34].

$$NA_{diode} = \sin\left(\frac{\theta}{2}\right)$$

(2.1)

28

In this Equation, θ is the full divergence angle of the laser source and NA_{diode} is the numerical aperture of the diode. Typically for an elliptical beam, the larger of the two angles (horizontal vs vertical) is used to compute the numerical aperture. The output beam diameter (ϕ) is chosen depending on the spot size that we would like to achieve with the collimated beam at the maximum range.

The goal now is to estimate the collimating lens parameters (focal length and numerical aperture) of the lens that are required to produce an output beam with the required diameter. Using simple trigonometry, the focal length of the lens is given by the following Equation (Equation 2.2).

$$\tan\left(\frac{\theta}{2}\right) \sim \frac{\phi}{2f_{lens}} \qquad (2.2)$$

As a rule of thumb, to capture the entire diverging beam from the laser source, the numerical aperture of the lens is typically set at twice the numerical aperture of the laser source [34]. Thus, the parameters of a collimator can be estimated given the size of the outgoing beam and the laser source numerical aperture. For example, given $\frac{\theta}{2} = 15°$ and the collimated beam diameter is 3 mm, the focal length of the lens is estimated to be 5.6mm and the NA of the lens is estimated to be 0.26 (Using Equation 2.1 and 2.2) [34]. The output beam diameter is chosen based on the required spot size, size of the scanning mirror used and the eye-safety considerations.

Before elaborating on the approach to compute the divergence of the outgoing beam, a few assumptions are made [35]:

i)   Thin lens approximation is valid – The thickness of the lens is negligible and does not contribute to the focal length of the lens.

ii)   Paraxial approximation is valid – Small angle assumption $\sin(\theta) \sim \theta$.

iii) Aberration free lens – No Chromatic or Spherical aberration.

iv)   Optical Invariant Equation is valid.

Figure below (Figure 9.3) is used as a reference to derive the basic optical invariant equation for a bi-convex lens. The result, however, applies to any type of lens. According to the optical invariant equation, the product of image size produced by a lens and ray angle is a constant.

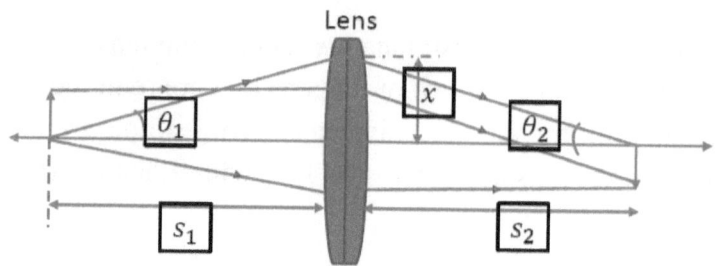

Figure 2.3 Optical Invariant – derivation for a generic bi-convex lens

Let an object be placed at a distance s_1 from the lens. The bottom of the object emanates a ray that intersects with the lens at a height x above the principle axis (OO') of the lens. The angle made by the ray with the optical axis is θ_1. The refracted ray intersects with the optical axis on the other side of the lens. The refracted ray makes an angle θ_2 with the optical axis. A ray from the top of the object, parallel to the lens refracts to pass through the focal point of the lens. The real image of the object is inverted and is as shown in the Figure above. Using simple trigonometry,

$$\theta_1 = \frac{x}{s_1}, \ \theta_2 = \frac{x}{s_2} \qquad (2.3(a))$$

$$\Rightarrow \theta_2 = \theta_1 \, x \frac{s_1}{s_2} \qquad (2.3(b))$$

The magnification of the lens is given by,

$$\frac{s_1}{s_2} = \frac{y_1}{y_2} \qquad (2.4)$$

$$\Rightarrow \theta_2 = \theta_1 \times \frac{y_1}{y_2} \tag{2.5}$$

$$\Rightarrow \theta_2 y_2 = \theta_1 y_1 \tag{2.6}$$

where y_1 is the object height, y_2 is the image height and s_2 is the image distance.

Equation 2.6 is the optical invariant equation. Let's considered another approach to estimate the focal length and divergence of the output beam. In the Figure 2.4, the lens parameters and the angles marked are self-explanatory. Using basic trigonometric rules and the small angle approximation, we have,

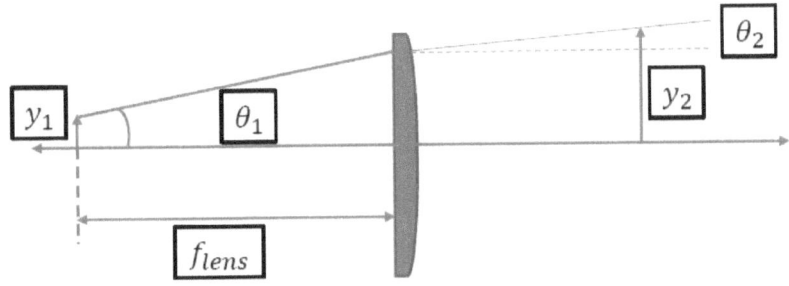

Figure 2.4 Collimation using a plano-convex lens

$$y_2 \sim f_{lens} \theta_1 \tag{2.7(a)}$$

$$f_{lens} \sim \frac{y_2}{\theta_1} \tag{2.7(b)}$$

$$Optical\ Invariant \rightarrow \theta_2 \sim \frac{y_1 \theta_1}{y_2} \tag{2.7(c)}$$

$$\Rightarrow \theta_2 = \frac{y_1}{f_{lens}} \tag{2.7(d)}$$

where y_1 is the radius of the fiber optic cable tip. Thus, the divergence angle (θ_2) and the focal length of the lens required to collimate the outgoing beam can be calculated using the equations above. Keeping the divergence angle to a minimum ensures that the beam is collimated for longer ranges and helps achieve better angular resolution for the lidar system. To achieve this a larger output beam diameter can be used.

The output beam from the laser collimator is typically referred to as a gaussian beam. This means that the irradiance profile of the beam is a gaussian function of range. A sample irradiance profile is given by Equation 2.8 below [26].

$$E(r,z) = E_o(z)e^{-\frac{2r^2}{w^2(z)}} \tag{2.8}$$

In the equation above, z is the range distance, w is the beam radius which is a function of z, and E_o is the peak irradiance at the center of the beam. The radius of the beam as it exits the collimator is known as the beam waist w_o. It is defined as the "$1/e^2$- width" of the beam irradiance profile. The collimator parameters chosen for the lidar prototype is listed in Table 2.1. A doublet achromatic lens is used for the collimator. A doublet lens uses two lenses cemented together to better handle aberrations that arise in a single lens. The lens ideally should improve optical throughput and correct any spherical or chromatic aberrations [36].

Table 2.1 Example Collimator Properties

SPECIFICATION	DESCRIPTION
Beam Diameter	7.0 mm
Divergence	0.016°
Numerical Aperture	0.24
Focal Length	37.13 mm
Lens Type, Anti-reflection coating	Air Spaced Doublet Lens, AR – C (type)

The optical beam from a point source emanates as a spherical wave characterized by its curvature. As it propagates along the range, its curvature increases and reaches a maximum i.e., becomes a planar wave with infinite curvature. The location where the curvature reaches its maximum is called the Rayleigh range z_R. It is given by the Equation 2.9 [26].

$$Z_R = \frac{\pi w_0^2}{\lambda} \tag{2.9}$$

The beam begins to diverge after the maximum curvature point. In the far-field i.e., $z > 5z_R$, divergence of the beam (θ_o) and laser beam spot size $(w(z))$ can be estimated using Equation 2.10 and 2.11[26].

$$\theta_o = \frac{\lambda}{\pi w_o} \tag{2.10}$$

$$w(z) = w_o \left(1 + \left(\frac{z\lambda}{\pi w_0^2}\right)^2\right)^{0.5} \tag{2.11}$$

where λ is the wavelength.

It is required that the optics used to collimate the beam is diffraction limited. A 'Diffraction Limited Beam' [37] is an ideal beam with no phase variations across the wave front thus helping us achieve diffraction limited divergence. Laser beams emanating from the collimator will by nature undergo diffraction. These wave fronts have a very strong spatial and temporal coherence, producing a collimated beam with minimum divergence in the far field. A non-ideal beam with phase fluctuations produces random spatial distributions and deviates from a gaussian beam.

2.5 LIDAR EQUATION
A pulse train is generated using the laser source and is amplified to obtain the required peak power for each pulse to be transmitted out into the atmosphere through a collimator. Laser beam irradiance is defined as the ratio of the peak power to the area of the laser beam spot at any given range. Thus, irradiance is a function of range distance. Area of the beam spot at any given range (R) can be calculated using the Equation 2.12 [38]. The equation uses the beam radius and the divergence angle (θ_{div}) of the beam to calculate the spot size (A_{spot}) at any given range distance. (Figure 2.5).

$$A_{spot} = \pi \left((\theta_{div}R) + \frac{\phi}{2}\right)^2 \tag{2.12}$$

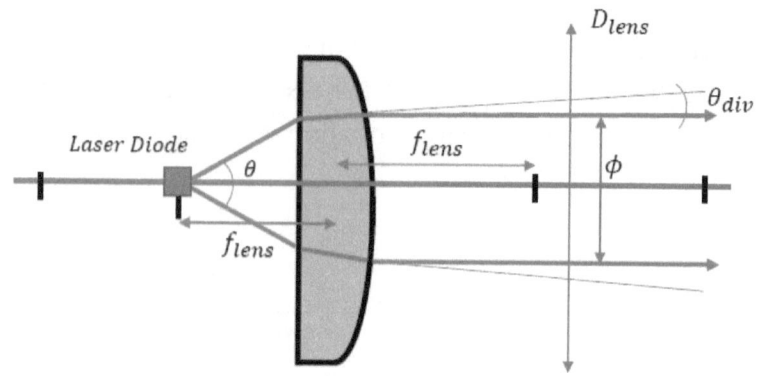

Figure 2.5 Plano-convex lens as a collimator with the output laser beam parameters.

The laser pulse as it propagates through the atmosphere is attenuated due to scattering off of random aerosol particles or absorbed by moisture/water molecules in its pathway. This results in lower optical power being transmitted and fewer photons reaching the target. The atmospheric transmission efficiency can be quantified using Beer's law (Equation 2.13) [26] [38]. The attenuation coefficient α is a function of range (R) and the wavelength (λ) of the laser. It is integrated over the range of interest to obtain the transmission efficiency. For lidars developed for automotive applications, the working range is within 200 m and the attenuation coefficient is assumed to vary linearly within this range. A typical value used in the literature is 0.12/Km [39]. The transmission efficiency ($\tau_{a,atm}$) is given by Equation 2.13(a),

$$\tau_{a,atm} = e^{-\alpha(\lambda)R} \qquad (2.13(a))$$

$$\alpha(\lambda) = \int_0^R \alpha(R)\,dR \qquad (2.13(b))$$

For the purposes of this discussion, the target is assumed to be at a range R. The transmission efficiency is assumed to be the same to and from the target. Irradiance (I_{laser}) of the beam at the target is given by

34

Equation 2.14. It accounts for the forward transmission efficiency parameter τ_a, which varies between 0 and 1.

$$I_{laser} = \frac{\tau_{a,atm} P_{peak}}{A_{spot}} \tag{2.14}$$

Another important aspect of the target is its reflectivity ρ, a measure of the ability of the target to reflect laser radiation. Quantitatively, reflectivity is the fraction of incident optical power that is scattered back by the target.

The target reflection can be one of diffuse, specular or retro reflection. If the target reflectivity is specular (mirror like reflection) the angle of the reflected beam is equal to its angle of incidence. Diffuse targets are commonly referred to as Lambertian surfaces. Reflection from a Lambertian target appears equally bright in all directions. If the incident beam is reflected along its same path of incidence, it is referred to as retro-reflection. Figure 2.6 [26] demonstrates the three reflection mechanisms discussed above.

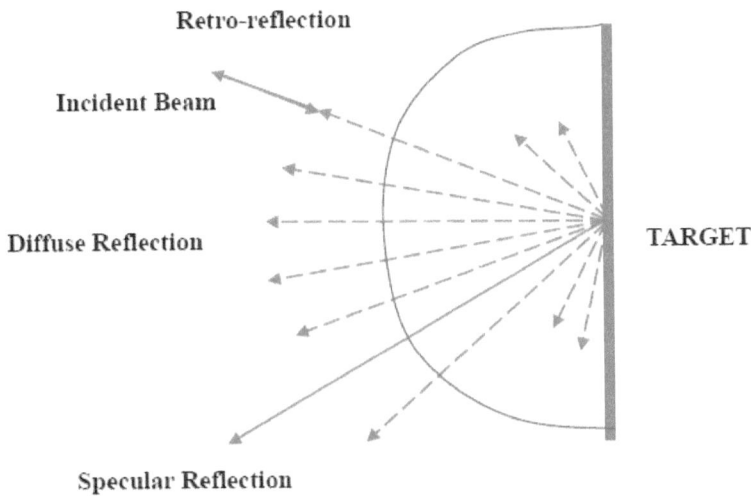

Figure 2.6 Target Reflection Mechanisms [26]

35

The brightness of the reflected radiation is defined as,

$$L_{target} = \frac{\rho\,\tau_a\,P_{peak}}{\theta_R A}$$ (2.15)

where θ_R is the solid angle into which the target reflects the incident radiation.

The solid angle is shown in Figure 2.7,

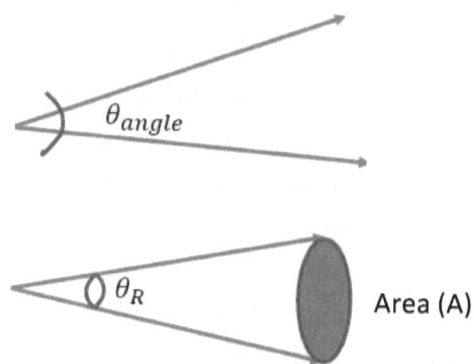

Figure 2.7 Angle between two lines (above) and solid angle subtended by a 2D area element (below).

$$\theta_R = \frac{A}{R^2}$$ (2.16)

where A is the area at a distance R.

For a Lambertian target this solid angle is π steradians, i.e., the reflected beam is scattered into a hemisphere that encompasses the entire target (since the projected angle of the hemisphere on the target is equal to π steradians) [36]. In Equation 2.15, A is the area of the surface that contributes to the return signal. A is either the area of the beam spot or the area of the target at the specific range. The minimum of the two limits the total returns collected at the receiver. Figures 2.8(a) and (b) demonstrate the two cases discussed.

Figure 2.8(a) Beam spot area > Target cross-section (Solid Block)

Figure 2.8(b) Beam spot area < Target cross-section (Solid Block)

Thus, the brightness of the backscatter from the Lambertian target is given by (Assuming '$A = A_{target}$'),

$$L_{target} = \frac{\rho \, \tau_a \, P_{peak}}{\theta_R \, A_{target}} \qquad (2.17)$$

A portion of the return signal is captured by the receiver. The irradiance of the return signal is given by,

$$I_{return} = \tau_a \times L_{target} \times \Omega_{FOV} \qquad (2.18)$$

where Ω_{FOV} is the solid angle subtended by the receiver (detector + lens) at the target.

Thus, the total return power (P_{return}) collected at the receiver is given by the product of the irradiance (I_{return}) and the area of the receiver lens ($A_{receiver}$). The complete equation (Equation 2.20) is known as the 'Lidar Equation' for a TOF optical sensor.

$$P_{return} = I_{return} \times A_{receiver} \qquad (2.19)$$

$$\Rightarrow P_{return} = \frac{\rho \, \tau_a^2 P_{peak} \Omega_{FOV} A_{receiver}}{\theta_R \, A_{target}} \qquad (2.20)$$

CHAPTER 3: TOF SYSTEM DESIGN - LIGHT RECEIVER

3.1 BACKGROUND POWER ESTIMATE – SOLAR RADIATION

Lidar sensors used under broad daylight conditions can collect solar photons scattered from the target or any surface within the detectors angular field of view. Using the model lidar equation, the total return power (from solar photons) collected by the receiver can be estimated. It is given by,

$$P_{background} = \tau_a \rho \, L_{solar,\lambda} \, A_{receiver} \, \Omega_{FOV} \, \Delta_\lambda \qquad (3.1)$$

where $L_{solar,\lambda}$ is the brightness of the solar radiation at a specific wavelength. $A_{rceiver}$ is the area of the receiver and Ω_{FOV} is the solid angle of the receiver subtended at the target.

The solar spectral irradiance ($I_{solar,\lambda}$, Figure 3.1) can be used to estimate the solar spectral radiance (brightness). Using the Lambertian target assumption and the standard irradiance values for solar radiations, the total amount of background power collected at the receiver can be estimated. Spectral bandwidth is denoted by Δ_λ. Solar radiation is inherently broadband, hence only photons within a bandwidth centered about the 1550 nm wavelength will be collected at the receiver. Here, only the one-way transmission efficiency parameter is used in the final expression.

$$I_{solar,\lambda} = \pi \, L_{solar,\lambda} \qquad (3.2)$$

$$P_{total} = P_{return,signal} + P_{background} \qquad (3.3)$$

where P_{total} is the total power collected at the receiver. $P_{return,signal}$ is the return power from the true signal and $P_{background}$ is the returns from the background radiations (solar photons).

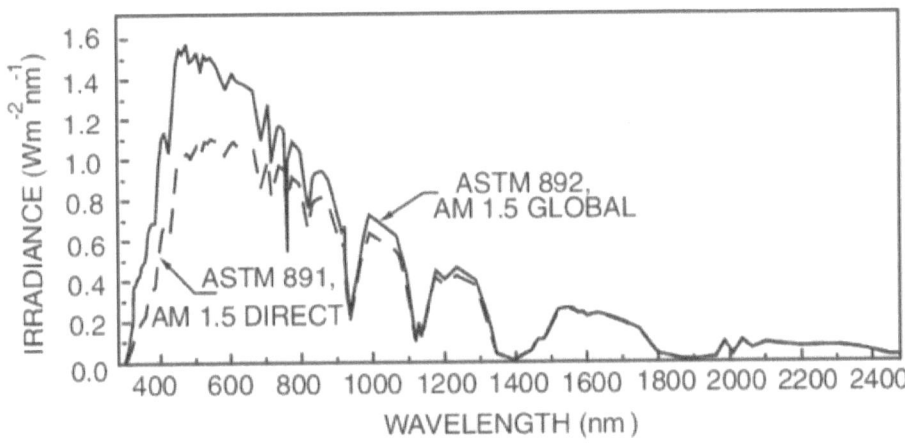

Figure 3.1 Solar Spectral Irradiance ASTM standard (average Conditions in the 48 states USA) [40]

3.2 RECEIVER OPTICS

This component of the lidar system constitutes one or more lenses, whose main functionalities are: a) To efficiently collect the returns from target b) To focus the return signal onto the detector surface. Typically, a lens is characterized by its focal length (f_{lens}) and diameter (D_{lens}). These parameters should be identified such that the lens performs each of the above tasks. Sometimes, the lens spec contains only one parameter, f-number, which is actually a combination of the above two parameters and is given by $f_\# = f_{lens}/D_{lens}$. A plano-convex lens is as an ideal solution for the receiver lens. Using the fundamentals of geometric optics, the returns can be tracked using the ray tracing technique as shown in the Figure 3.2.

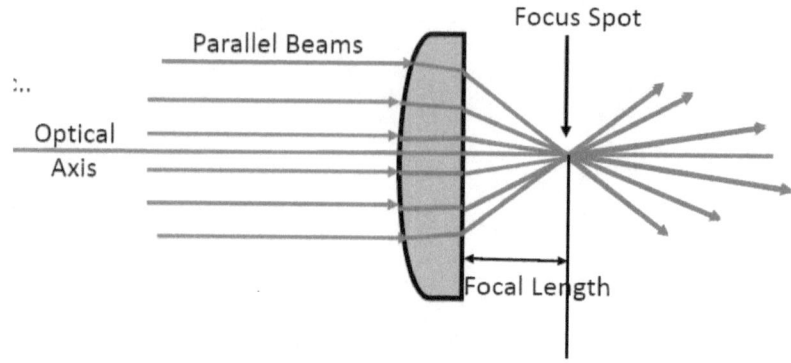

Figure 3.2 Ray tracing technique to image the rays from a distant target using a Plano-convex lens.

There are two types of lidar transmitter/receiver (transceiver) systems. One is the mono-static system and the other is a bi-static system [26] [38].

In a mono-static system, the transmitter optics and the receiver optics are one and the same (Figure 3.3). A special transceiver switch (typically a quarter wave plate) is used to isolate the transmitted and received beams. This is paramount to ensure that the detector does not pick up photons from the transmitted beam and trigger a false positive event, thereby introducing inaccuracies in the TOF estimate. A mono-static system requires additional optics and high precision alignment of these optical components for effective performance. Hence, it is relatively more expensive and complicated when compared to a bi-static system. However, as the mono-static system uses the same lens to transmit and capture the returns from the target, the amount of background photons picked up by the sensor is much less when compared to a bi-static system.

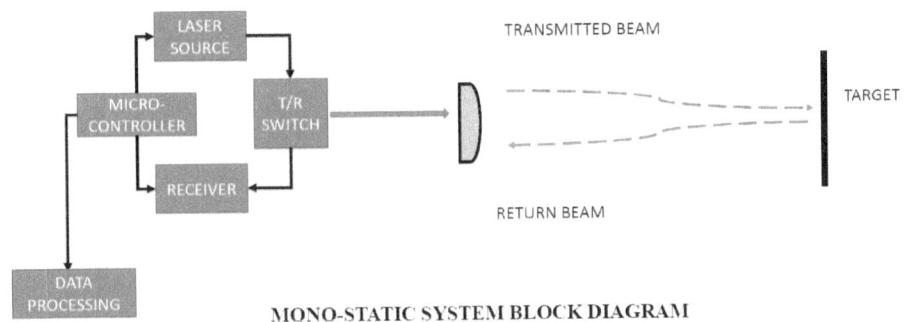

MONO-STATIC SYSTEM BLOCK DIAGRAM

Figure 3.3 Mono-static system block diagram.

A bi-static system on the other hand, uses two separate channels for the transmitted and received optical signals. Figure below describes a bi-static system (Figure 3.4). A lidar with a bi-static configuration is easier to build and has fewer optical components that make the system relatively inexpensive.

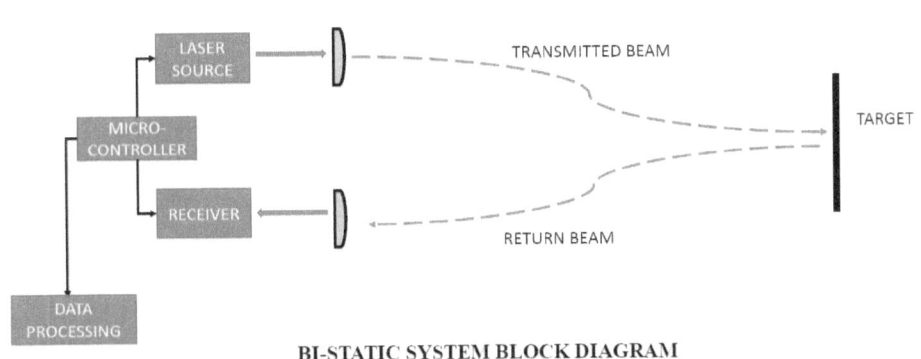

BI-STATIC SYSTEM BLOCK DIAGRAM

Figure 3.4 Bi-static system Block Diagram

Keeping the transmitted and received signal pathways independent avoids false triggering due to the transmitted pulses. However, a bi-static system inherently has a blind zone in front of the lidar sensor. There is a region in front of the sensor as show in Figure 3.5 [41] [42] that doesn't fall within the receiver lens FOV. Returns from this region (blind zone) will not be captured by the receiver lens and hence targets in the blind zone will not be detected.

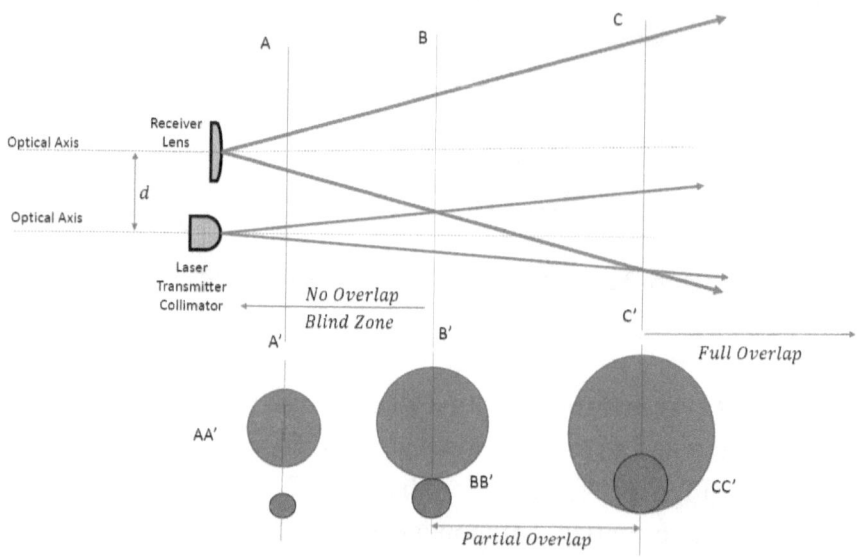

Figure 3.5 Blind zone, region of partial overlap and full overlap for a bi-static system [41].

Knowing the distance to the partial overlap (BB' in the above figure) would help us understand the minimum distance at which the sensor starts detecting a target. Ideally, to get a strong detection, we would require the target to be beyond the full overlap point (CC' in the above figure). Distance to BB' and CC' can be estimated using the equations derived later Equations 3.7(c) and 3.8(b).

If the target is not in the blind zone region, the receiver will be able to focus the return signal photons on to the detector. The detector is typically a single pixel photodiode that is positioned at the focal point of the lens. In addition to being outside the blind zone, the target should be within the angular FOV of the detector.

Figure 3.6 describes the FOV concept for the receiver sensor. The characteristic length of a single pixel detector (size of the detector) projected through the lens sets the angular FOV for the receiver (β in Figure 3.6). If the target is within the FOV the returns will be focused onto the detector. The angular FOV is estimated using Equation 3.4. Target within this angular FOV will be detected by the sensor [43].

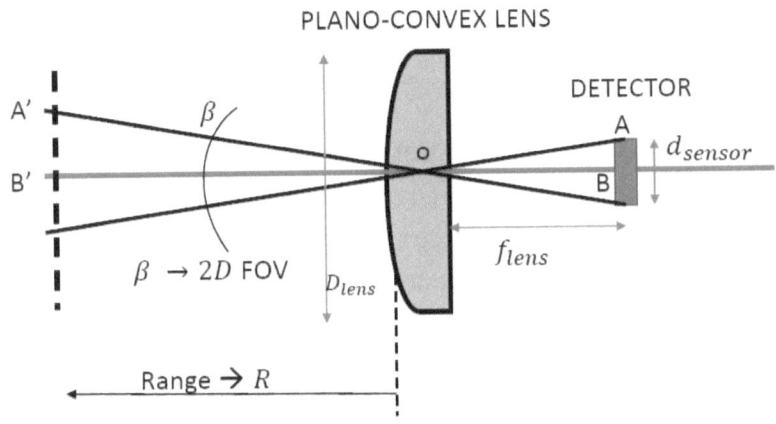

Figure 3.6 Angular FOV of a lidar receiver (lens + detector) system

In Figure 3.6 consider triangles oAB and triangle oA'B'. Using the law of similar triangles,

$$\tan\left(\frac{\beta}{2}\right) = \frac{d_{sensor}}{2 f_{lens}} = \frac{D_{FOV}}{2R} \tag{3.4}$$

In the Equation 3.4, β is the angular field of view for a given detector size d_{sensor}, and a plano-convex lens of focal length f_{lens}. Here D_{FOV} is the instantaneous field of view (IFOV), the projected dimension of the sensor in the scene ahead. The IFOV is directly proportional to the of distance from the sensor (Equation 3.5).

$$\Rightarrow D_{FOV} = \frac{d_{sensor} \, R}{f_{lens}} \tag{3.5}$$

For a given range, the detector is capable of capturing photons from the projected area (A_{FOV}). It is calculated using the Equation 3.6(a) or (b) [43].

For a square detector,

$$A_{FOV} = D_{FOV}^2 \qquad\qquad (3.6(a))$$

For a circular detector,

$$A_{FOV} = \frac{\pi D_{FOV}^2}{4} \qquad\qquad (3.6(b))$$

Referring back to the lidar equation (Equation 2.20, rewritten here for convenience), the target area (A) is the minimum of beam spot area and target area. Based on the discussion above the projected area has a role to play in deciding the target area, in addition to the spot and target areas.

$$\Rightarrow P_{return} = \frac{\rho\, \tau_a^2 P_{peak} \Omega_{FOV} A_{receiver}}{\theta_R\, A}$$

Hence $A = \min(A_{spot}, A_{target}, A_{FOV})$

The range distance over which a bi-static system has complete overlap between the receiver FOV and the laser beam, can be calculated using simple trigonometry as shown below. Figure 3.7 shows the basic setup for the calculation. The separation between the collimator lens and the receiver lens is d (cm). The laser beam has a divergence given by θ_{div} and the angular FOV of the receiver lens is β and the half angle is '$\theta_{FOV} = \beta/2$'

Using Figure 3.7 [42],

$$\tan(\theta_{FOV}) = \frac{d - r_b(x_{min})}{x_{min}} = \frac{d - x_{min}\tan(\theta_{div})}{x_{min}} \qquad\qquad (3.7(a))$$

$$\text{Using } r_b(x_{min}) = x_{min}\tan(\theta_{div}) \qquad\qquad (3.7(b))$$

$$\rightarrow x_{min} = \frac{d}{\tan(\theta_{FOV}) + \tan(\theta_{div})} \qquad\qquad (3.7(c))$$

Similarly,

$$\tan(\theta_{FOV}) = \frac{d + r_b(x_{max})}{x_{max}} = \frac{(d + x_{max}\tan(\theta_{div}))}{x_{max}} \quad (3.8(a))$$

$$\rightarrow x_{max} = \frac{d}{\tan(\theta_{FOV}) - \tan(\theta_b)} \quad (3.8(b))$$

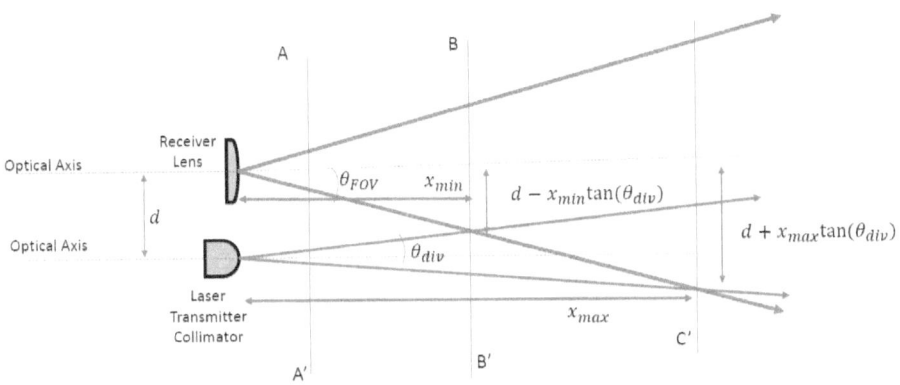

Figure 3.7 Partial Overlap and Full Overlap range distance calculation.

For the given lidar prototype parameters (Table 3.1) x_{min} and x_{max} as shown in Figure 3.8 and 3.9.

Table 3.1 Lidar Prototype Parameters

LIDAR PROTOTYPE PARAMETER	VALUE
Beam Divergence	0.016°
Receiver FOV	2.2546°
Sensor Diameter	2 mm
Focal Length	25.4 mm

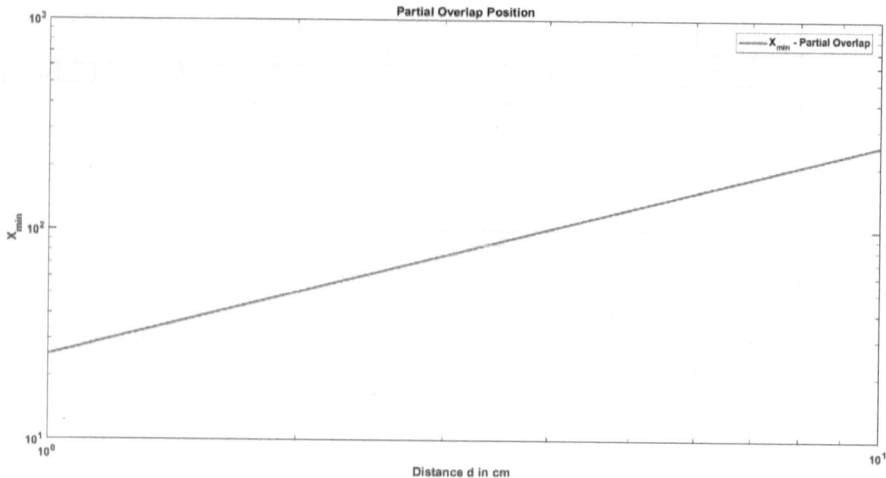

Figure 3.8 Partial overlap range distance as a function of separation distance between the receiver and the transmitter optics.

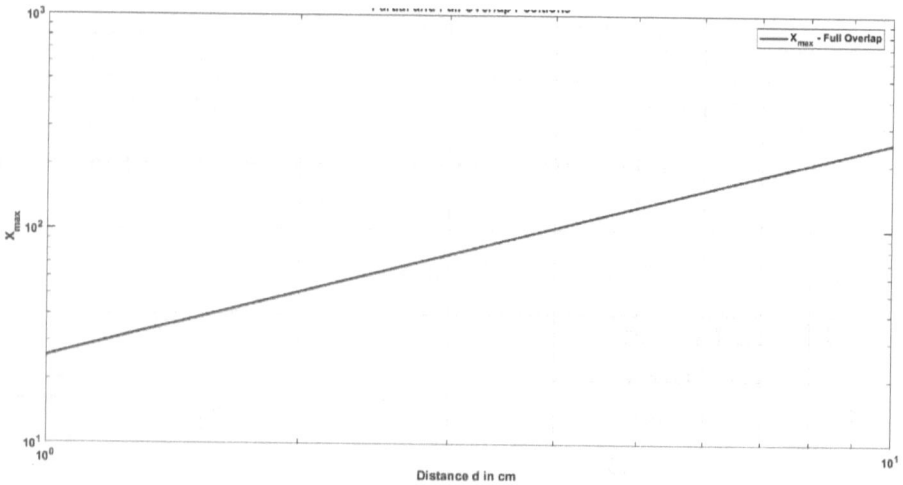

Figure 3.9 Full overlap range distance as a function of separation distance between the receiver and the transmitter optics.

46

3.3 RECEIVER LENS PARAMETERS

The linear relationship between the return power collected by the receiver and the receiver lens area is evident from the lidar equation derived previously (Equation 2.20). Using a larger diameter lens will increase the number of backscattered photons collected from the target. It also results in more background solar photons to be collected, thereby making the true return signal noisier.

$$P_{return} \propto A_{reciever} \Omega_{FOV} \qquad (3.9)$$

As we already know, the receiver lens has the added responsibility to focus the return signal back onto the detector. The spot size produced by the lens is required to be smaller than the detector pixel element size. Using techniques from geometrical optics, the effects of the lens parameters on the spot size produced by the lens can be studied.

The focusing capability of the lens is characterized by the spot size produced by the lens. In Figure 3.10, for an extended source of height y_1 located (on-axis) at a distance s_1 ($2f_{lens} < s_1 \leq \infty$), an image with height y_2 is formed behind the lens - closer to its focal point. Based on the fundamental imaging principles of a lens, objects at distance $> 2f_{lens}$ will form an image between f_{lens} and $2f_{lense}$ (behind the lens). With the object moving closer to infinity, the image moves closer to the focal point behind the lens.

Now consider, the beam spot area at a given range. The radius of this spot is equal to y_1 in Figure 3.10. The rays emanating from the spot, focused by refraction, intersect the focal plane (the plane perpendicular to the optical axis that contains the focal point) and produce an image behind the focal point (distance s_2) with height y_2.,

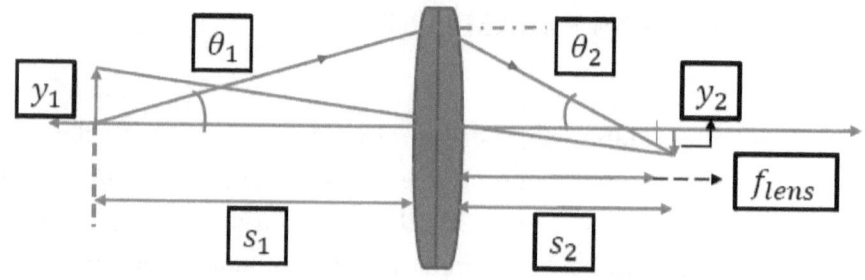

Figure 3.10 Lens spot size for an on-axis extended source

Using the paraxial approximation and the optical invariant theorem, y_2 is given by Equation 3.12.

For $s_1 \gg 2f_{lens}$ → Image forms at the focus

For small angles θ_1, θ_2 à $\theta_2 = \dfrac{R_{lens}}{f_{lens}}$ $\qquad$ (3.10)

Invariant theorem → $y_1\theta_1 = y_2\theta_2$

$$\Rightarrow y_2 = \left(\frac{f_{lens}}{R_{lens}}\right)y_1\theta_1$$

$$\Rightarrow y_2 = \left(\frac{f_{lens}}{2R_{lens}}\right)(2y_1)\left(\frac{R_{lens}}{s_1}\right) \qquad (3.11)$$

$$y_2 \sim 2y_1\left(\frac{R_{lens}}{s_1}\right)f^{\#} \qquad (3.12)$$

where $f^{\#}$ is the f-number of the lens i.e., ratio of the focal length of the lens and the diameter of the lens, R_{lens} is the lens radius [44].

From the equation above (Equation 3.12), to obtain a tight image spot i.e., to focus the beam to a tight spot, the f-number of the lens should be small. Commonly available lenses from Thorlabs, Newport etc. have f-number as small as 1, i.e., $f_{lens} = D_{lens}$. For a fixed lens radius (R_{lens}), a smaller focal length would produce a tighter spot. For targets that are far away ($s_1 \gg 2f_{lens}$) the spot size is small. Size of the extended source also influences the spot size. A smaller beam spot at the target would aid the lens in focusing the rays from the target to a tight spot behind the lens.

The f-number of a lens can be reduced by using an aperture stop that reduces the diameter of the lens. This reduces the f-number and helps focusing the returns to a tighter spot [44]. However, the aperture stop reduces the effective diameter of the lens, thereby lowering the return power collected by the lens. There is a trade-off between receiver collection efficiency and spot size reduction. Figure 3.11 and Figure 3.12 demonstrate the effect of the detector size and focal length on the lidar receiver field of view (FOV). For a fixed focal length of the lens a detector with larger area will have a larger angular field of view. A larger angular field of view ensures that the range distance of full overlap is closer to the lidar sensor. Thus, reducing the blind zone region for a given lidar prototype.

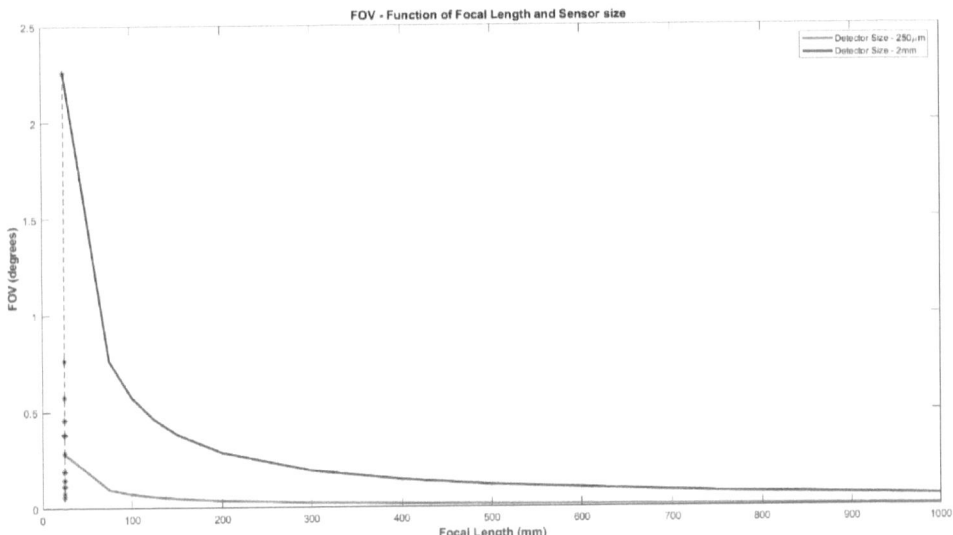

Figure 3.11 FOV of the receiver as a function of focal length of the lens and sensor size (*--*-- → prototype focal length 37.4 mm).

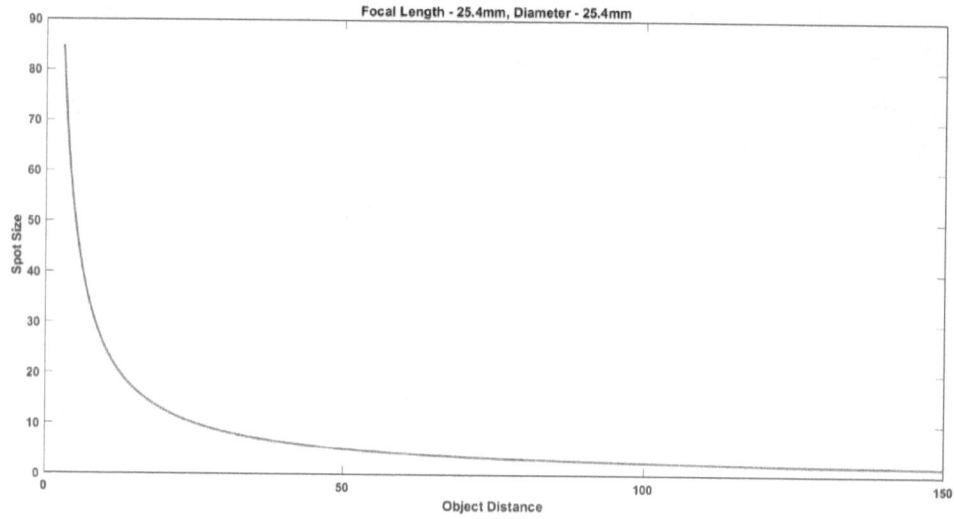

Figure 3.12 Receiver lens spot size (in mm) for the lidar prototype as a function of range (in mm).

An important question arises at this point: how far is far enough to assume that the angles made by the incident rays are small enough? (Figure 3.13). To analyze this scenario, consider Figure 3.13.

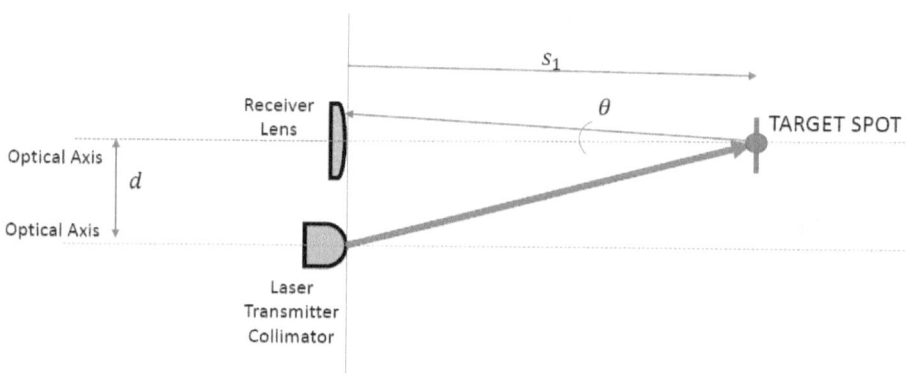

Figure 3.13 On axis target focusing using a plano-convex receiver lens.

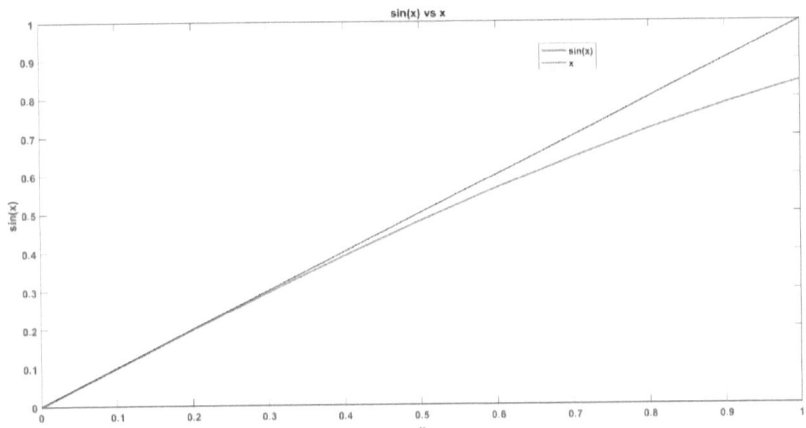

Figure 3.14 Small angle approximation Sin(x) vs x (x in radians)

The largest angle formed between the lens and the rays emanating from the target spot (as shown in the Figure 3.13), θ is given by,

$$\theta = \tan^{-1}\left(\frac{R_{lens}}{s_1}\right) \tag{3.13}$$

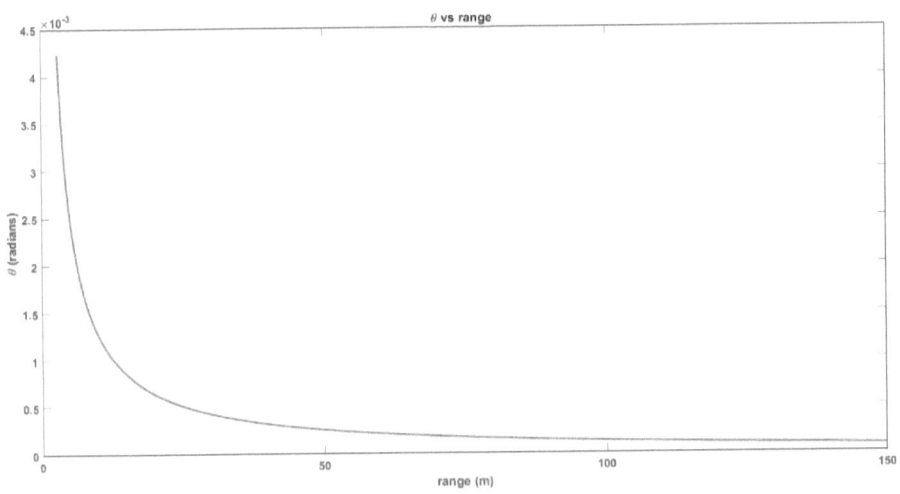

Figure 3.15 θ **vs range distance for the lidar example – validity of small angle approximation.**

Consider a lidar example designed as a bi-static system that can range targets between 3 meters and 150 meters. For a 1" diameter lens, Figure 3.15 (above) describes the change in θ as a function of range. We can see that the maximum value for θ is 0.004 radians. From Figure 3.14 we can see that for angles $< 0.4\ radians$, small angle approximation holds true. Thus, the approximation holds true for this example (on-axis target) as well.

For an off-axis target as shown in Figure 3.16, a similar analysis can be repeated to ensure that the small angle assumption holds true for the range of operation (3-150 m).

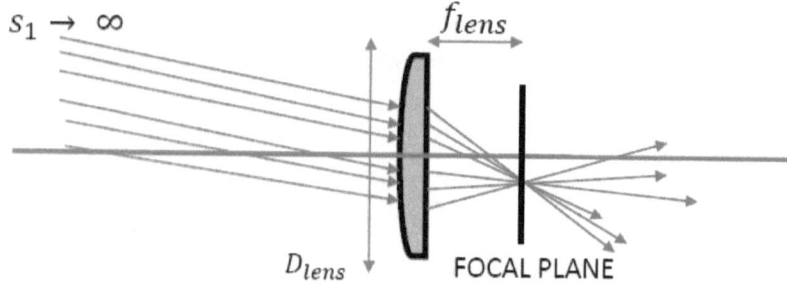

Figure 3.16(a) Focusing off-axis parallel beams using a plano-convex lens.

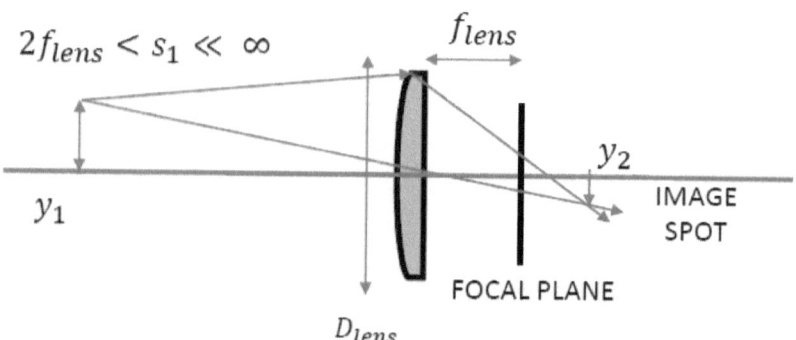

Figure 3.16(b) Focusing non-parallel rays from off-axis target spots using a plano-convex lens.

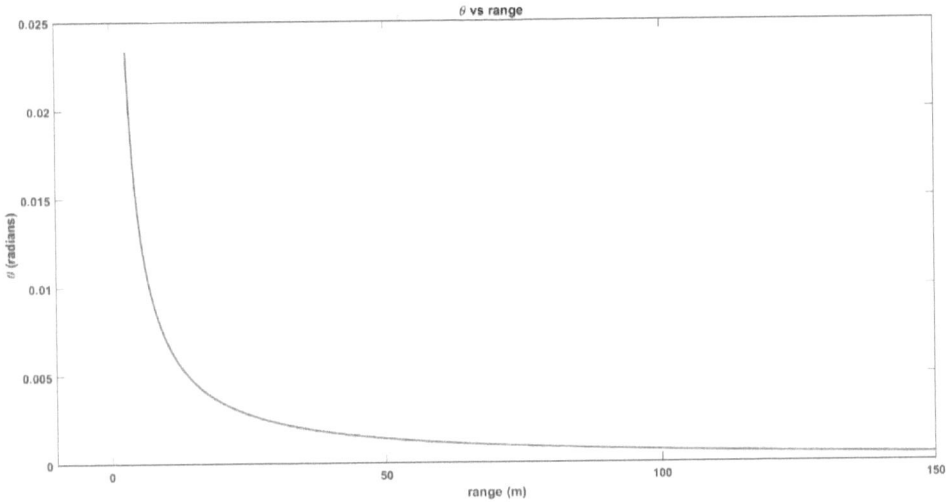

Figure 3.17 θ vs range for an off-axis target.

From Figure 3.17 the assumption holds true for the off-axis case as well. Size of the detector should be such that the returns from the receiver lens are focused efficiently onto the detector surface. The size of the focused spot produced by the receiver lens can be estimated. Figure below (Figure 3.18) shows a plano-convex lens and how the rays emanating from an off-axis target is focused by the lens. The lens is assumed to be a thin lens and the optical invariant equation will be used in the following derivation [45].

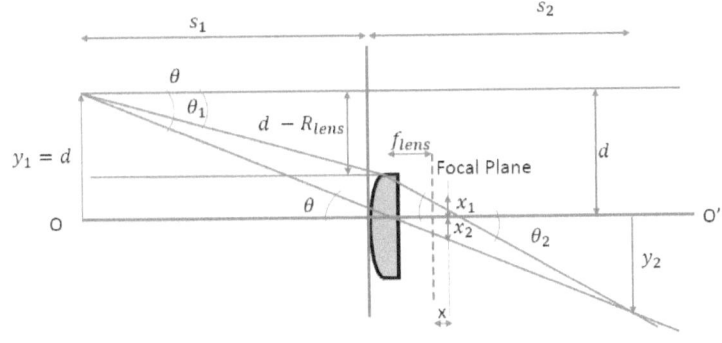

Figure 3.18 Focusing an off-axis beam using a plano-convex lens.

Starting with the known equations,

$$y_1 \theta_1 = y_2 \theta_2$$

$$Magnification \rightarrow M = -\frac{s_2}{s_1} \tag{3.13}$$

$$Lens\ Makers\ Equation \rightarrow \frac{1}{f_{lens}} = \frac{1}{s_1} + \frac{1}{s_2} \tag{3.14}$$

$$\tan(\theta_1) = \frac{d-R_{lens}}{L} \rightarrow \theta_1 = \tan^{-1}\left(\frac{(d-R_{lens})}{L}\right) \tag{3.15}$$

$$\tan(\theta) = \frac{d}{L} \rightarrow \theta = \tan^{-1}\left(\frac{d}{L}\right) \tag{3.16}$$

Using vertically opposite angle rule,

$$\tan(\theta) = \frac{x_2}{f_{lens}} \tag{3.17}$$

$$\rightarrow x_2 = f_{lens} \tan(\theta) \tag{3.18}$$

Using the lens maker equation, calculate s_2,

$$\frac{1}{s_2} = \frac{1}{f} - \frac{1}{s_1}$$

$$\tan(\theta_2) = \frac{y_2}{(s_2 - f_{lens}) - x} \tag{3.19}$$

$$\rightarrow x = (s_2 - f_{lens}) - \frac{y_2}{\tan(\theta_2)} \tag{3.20}$$

$$Also,\ \tan(\theta_2) = \frac{x_1}{x} \tag{3.21}$$

$$\rightarrow x_1 = x \tan(\theta_2) \tag{3.22}$$

The spot size above and below the optical axis can be estimated using Equations 3.18 and 3.22. Figure 3.19 plots the spot size as a function of range.

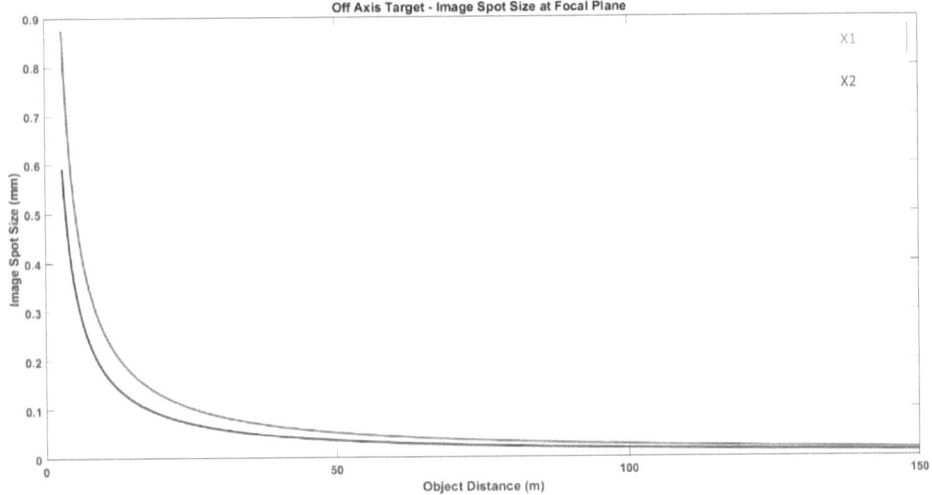

Figure 3.19 Spot size as a function of range $x_1 (Above), x_2 (Below)\ vs\ range$

The spot size is given by the sum of $x_1\ and\ x_2$. Given that the size of the detector for the lidar prototype is 2mm, Figure 3.19 shows that the spot is always within the detector and can be efficiently collected and focused by the receiver lens.

A more rigorous approach to estimate the spot size produced by a lens is by using a thick lens model [46]. Figure 3.20 shows a bi-convex thick lens with all the major definitions for a thick lens marked. The focal length of a thick lens is dependent on the thickness of the lens. The focal length is given by,

$$\frac{1}{f} = (n-1)\left(\frac{1}{R_1} - \frac{1}{R_2} + \left(\frac{(n-1)d_{lens}}{nR_1R_2}\right)\right) \qquad (3.23)$$

$$h_1 = -\frac{f_{lens}(n-1)d_{lens}}{R_2 n} \qquad (3.24)$$

$$h_2 = -\frac{f_{lens}(n-1)d_{lens}}{R_1 n} \qquad (3.25)$$

where n is the refractive index of the lens.

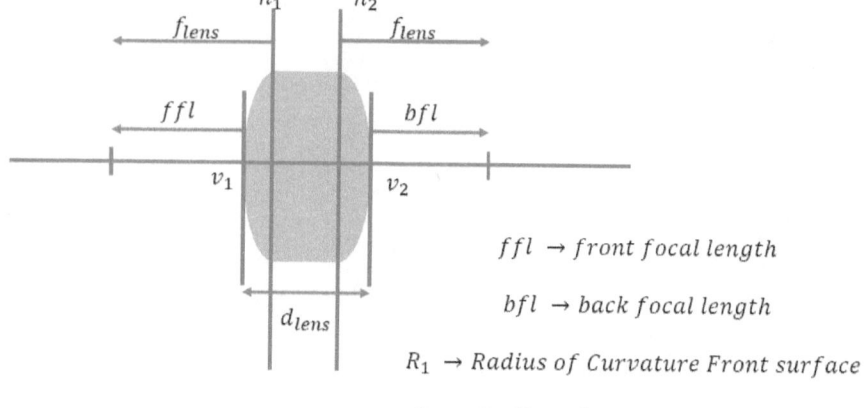

ffl → front focal length

bfl → back focal length

R_1 → Radius of Curvature Front surface

R_2 → Radius of Curvature Back surface

Figure 3.20 Thick lens parameters.

We at ASU developed a *Matlab*™ routine to setup a plano-convex lens (lens used for the lidar prototype) to simulate rays entering and exiting the lens randomly. We estimated the final output ray direction through the planar surface onto the focal plane using the 'Matrix Method'. For a thick lens, the incoming ray undergoes refraction at the front (curved) surface, then it translates through the lens and finally undergoes another refraction at the exit face.

The matrix method equation to trace the refracted ray for any lens is,

$$\eta_t = R_2'\ T_{21}\ R_1'\ \eta_i \qquad (3.26)$$

where,

$$R_2' = \begin{bmatrix} 1 & -D_2 \\ 0 & 1 \end{bmatrix}, R_1' = \begin{bmatrix} 1 & -D_1 \\ 0 & 1 \end{bmatrix}, T_{21} = \begin{bmatrix} 1 & 0 \\ \dfrac{d_{21}}{n_{t1}} & 1 \end{bmatrix}$$

$$\eta_t = \begin{bmatrix} n_{t2}\alpha_{t2} \\ y_{t2} \end{bmatrix}, \eta_i = \begin{bmatrix} n_{i1}\alpha_{i1} \\ y_{i1} \end{bmatrix}$$

$$D_1 = \frac{n_{t1} - n_{i1}}{R_1}$$

$$D_2 = \frac{n_{t1} - n_{i1}}{R_2} \qquad (3.27)$$

In the equation above, n_{i1} is the refractive index of the medium from which the ray is input, n_{t1} is the refractive index of the lens and n_{i2} is the refractive index of the exit medium. The matrices R_1', R_2' and T_{21} together constitute the transfer function. Each matrix calculates the ray angle and the ray height above the optical axis. In the equations above, α_{i1} is the angle of incidence of the ray and y_{i1} is the height of the ray above the optical axis. Also, α_{t2} and y_{t2} are the exit ray angle and height above the optical axis respectively. d_{21} is the thickness of the lens previously referred to as d_{lens}. The parameters for the lidar prototype plano-convex lens is preset for the simulations. The preset values are listed in Table 3.2.

Table 3.2 Plano Convex Lens – Thick Lens Ray Tracing Simulation Parameters.

PARAMETER	SPECIFICATION
d_{lens}	1.17 cm
$\phi_{diameter}$	2.54 cm
n_{t1} – refractive index	1.3
n_{i1} –refractive index	1
n_{t2} –refractive index	1
n_{i2} –refractive index	1.3
f_{lens}	4.3667 cm
R_1 – radius of curvature	1.31 cm
R_2 –radius of curvature	∞
h_1 – front principal plane	0
h_2 –back principal plane	-0.90 cm
ffl – front focal length	4.3667 cm
bfl –back focal length	3.4667 cm

For each ray, the height above the optical axis at which the ray intersects with the focal plane is estimated. A 10000-ray simulation was used to identify the distribution of these points behind the lens on its focal plane (Figure 3.21). The spot center can be estimated as the mean of this distribution and the diameter of the spot is twice the standard deviation of the distribution.

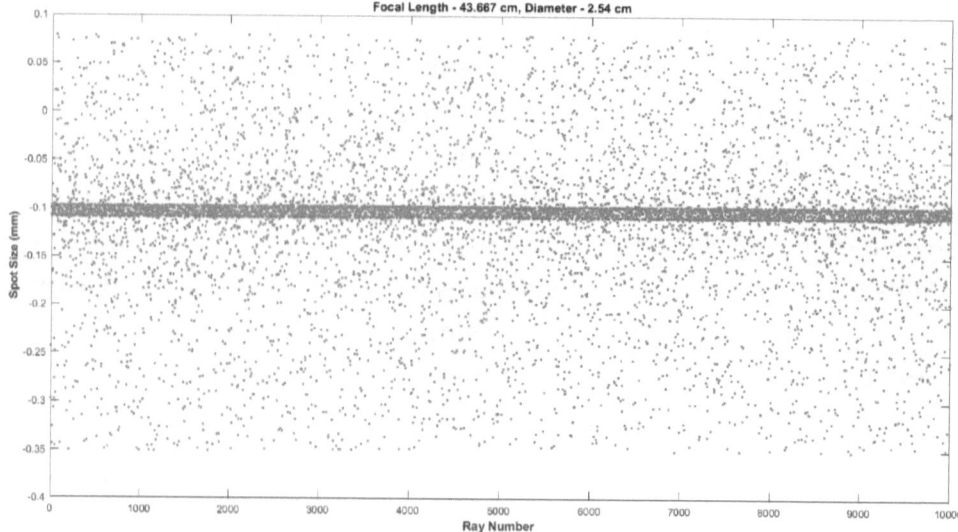

Figure 3.21 Spot size distribution. Range – 3 m, detector/collimator separation – 7cm.

For a target at 3 meters from a bi-static lidar system with 7 cm separation between the collimator and the receiver lens, the spot size (mean) estimated is 1.21 mm. The diameter of the spot is estimated to be 1.4 mm. The overlap factor for the detector can be calculated as the ratio of the overlap between the detector and the image spot area to the area of the detector. In general, the overlap factor (O.F.) (Figure 3.22) increases with range. Thus, only a percentage of the total returns P_{return} (return power collected by the receiver lens) based on the lidar equation will be focused on the detector. For the lidar prototype, the above setting will help successfully estimate the total returns from targets within the lidar working range, i.e., between 3 meters and 150 meters.

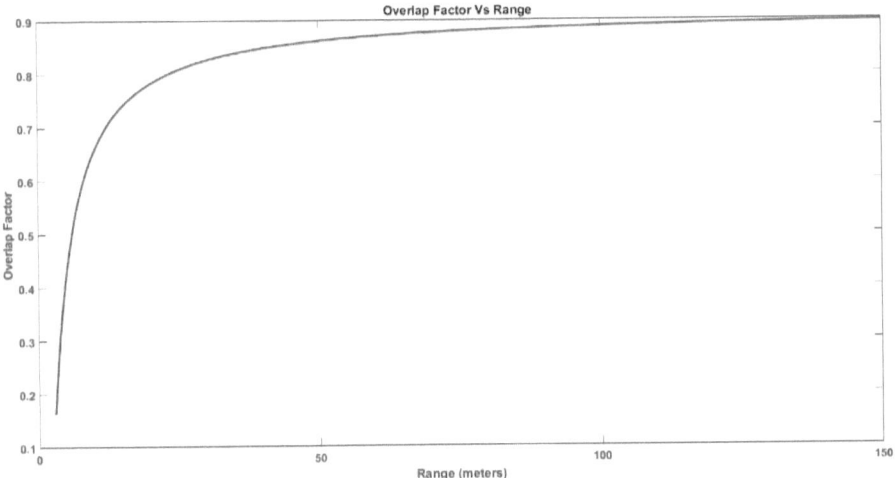

Figure 3.22 Spot Size Distribution. Range – 3 – 150 m, detector/collimator separation – 7cm.

CHAPTER 4: TOF SYSTEM DESIGN: PHOTONS TO ELECTRONS

4.1 PHOTODECTOR

The backscattered signal from the target is collected at the receiver and focused onto the detector surface using a focusing lens. The photodetector is the most important component in the lidar that helps us detect a target and estimate its depth information from the backscatter returns. The photodetector responds to the scattered photons and produces current (A) or voltage (V) proportional to the incident intensity of light. The choice of a detector is based on the operating wavelength of the laser device, the sampling speed or the bandwidth required, the maximum range of operation of the lidar system and (most importantly) the noise characteristics of the detector. In this section, we explore each facet in detail and elaborate on the limits set by a chosen detector on the sensing operation.

Lidar sensors typically use a single pixel photodetector to detect the return signal. An array of pixel is more common in applications where returns need to be sampled spatially and can be used to assign an angle of arrival for each individual line of sight return. A flash lidar or a simple scanning lidar is used widely in autonomous cars to function as their eyes (perception sensor). It uses an array of 16 to 32 detectors to sample the entire 360° field of view around the car.

4.1.1 MODES OF OPERATION

Photodetector is a simple semiconductor diode that exhibits a change in its characteristic when a photon strikes its surface. Incident photons produce electron-hole pairs within the semiconductor. The flow of these electrons produces a photocurrent equivalent to the strength of the incident light intensity.

Photodetectors are broadly classified based on their modes of operation [48] [49]. If the generated free electrons are guided by an applied electric field/potential (Reverse Bias Mode), the detector is called a **Photoconductive** detector. If no external bias (No Bias Mode) is applied the detector is called a **Photovoltaic** detector. A detector that emits free electrons from its surface due to an incident photon is called a **Photo-emissive** detector. The process of photoemission is fundamental to how a solar cell works to harness the sun's energy. The choice of the operating mode is based on our sensing application and a thorough evaluation of the different noise sources that contaminate the detector output signal. Quantifying the different noise sources that contaminate the signal will help us reduce uncertainty in detecting the true return signal. [50] [51].

4.1.2 DETECTOR CHARACTERISTICS

Incoming photon energy must be above the threshold limit [52] (Semiconductor Band Gap) of the semiconductor material to generate an equivalent photocurrent. Energy of a photon is characterized by its wavelength. So every detector has a specific response curve that characterizes the output of the detector as a function of wavelength. The curve helps us choose the right wavelength for our application. It ensures a strong detection probability when working within the range of frequencies to which the detectors can respond. For example, the Indium Gallium and Arsenic (InGaAs) based detector is suitable for applications that use wavelengths in the far infra-red (FIR) regime of the electromagnetic spectrum, i.e., wavelength greater than $1.4 - 2.0 \ \mu m$. We shall take a look at the important detector characteristics in the following sections. [55] [56]

4.1.3 RESPONSIVITY

Responsivity of a detector is defined as output of the detector per unit watt of input optical power. Its units are either Amperes/Watt or Volts/Watt, depending on whether the output from the detector is measured in amperes or in volts. It is a function of the wavelength of the received signal and is defined as,

$$R(\lambda) = \frac{e\eta}{h\nu} \qquad (4.1)$$

where e is the charge in an electron, η is the quantum efficiency, h is the Plank's constant and ν is the frequency of the photon.

4.1.4 QUANTUM EFFICIENCY

Quantum efficiency of a detector is the ratio of the number of electron-hole pairs generated for each incident photon. It is a measure of how effectively the incident photon/optical energy is converted to an output electric current or voltage. It is defined as,

$$\eta = \frac{n_e}{n_p} \qquad (4.2)$$

where n_e is the number of electrons emitted and n_p is the number of photons incident on the detector.

4.1.5 DETECTOR BANDWIDTH - RESPONSE TIME

Every detector has an intrinsic capacitance (junction capacitance C) and a load resistance R. Typically, a detector responds to the incident intensity of light. Detector bandwidth is a direct measure of how fast the detector can respond to fluctuations in intensity. Bandwidth (f_{BW}) of a detector [49] and the rise time t_r are defined by,

$$f_{BW} = 1/(2\pi RC) \qquad (4.3)$$

$$t_r = \frac{0.35}{f_{BW}} \qquad (4.4)$$

The product of R and C is termed as the time constant τ of the detector. It is a measure of how quickly the detector output decays after a light source instantaneously illuminates the detector. A small value of τ results in faster decay and sets the bandwidth of the detector to a high value, i.e., it is a very fast detector. A large time constant would result in a low bandwidth and hence a slow detector. The rise time t_r , is the amount of time that the output of the detector takes to rise from 10% to 90% of its peak output. The rise time metric is related to the bandwidth as shown in the Equation 4.4.

4.1.6 LINEARITY

The output of every photodetector has a lower limit and an upper limit of operation. Lower limit is set by the inherent noise characteristics of the detector. Upper limit, also known as the saturation limit of the detector, is the point beyond which there is no corresponding increase in the output of the detector with increasing input optical power. The detector needs to be operated within this limit to ensure accurate performance. The appropriate region of operation is when there is a linear relationship between the input optical signal power and the detector power output.

4.1.7 SIGNAL TO NOISE RATIO (SNR)

Signal to Noise Ratio or SNR is a metric used to measure the relative strength of signal in background noise. It is defined as the ratio of signal current (I_s) to noise current (I_n) measurements. Equation 4.5 is the SNR in terms of signal and noise current (SNR_c).

$$SNR_c = \frac{I_s}{I_n} \tag{4.5}$$

Incident signal photons accumulated (time-integrated) over a time duration τ_p, produce electron-hole pairs in the detector. Any source that contributes to the generation of these electron-hole pairs in the detector material apart from the true signal photon is considered a noise source. The most common sources of noise in a detector that contributes to the noise current are, dark current (i_d), thermal noise (i_{th}) and shot noise (i_s).

$$I_n = i_d + i_{th} + i_s \tag{4.6}$$

The dark current or leakage current noise (i_d) in a detector arises due to the external bias applied (photoconductive mode) on the detector. When operating in photovoltaic mode, the strength of the dark current is kept to a bare minimum.

Thermal noise or 'Johnson Noise' produces free electrons in the detector due to random temperature fluctuations. Contribution to the noise current due to these random fluctuations is quite small (i_{th}). By operating the detector and the entire electronic circuitry associated with the detector in a temperature-controlled environment the effect of thermal noise can be kept under control.

Shot noise is the predominant source of noise within a photodetector that arises due the discrete nature of the incident photons and emitted electrons. Shot noise produces a time dependent fluctuation in the output current produced from the detector. All other sources of noise are considered negligible. Lidar systems that use such a detector are referred to as shot noise limited systems.

The number of electrons emitted over the duration τ_p is modeled based on a simple Poisson distribution statistic [54] and expressions for each noise source contribution can be derived based on their source of generation. The following equations are used widely to model the noise contributions due to each noise source. In the following equations, I_d is the mean dark current flowing through the circuit, I_s is the mean detector current produced as a response to the incident signal photons, and R is the resistance in the detector circuit giving rise to noise [53]. The noise contribution is bandwidth dependent and hence, larger the bandwidth, larger the noise contribution to the true signal.

$$i_d = (2eI_aB)^{\frac{1}{2}} \tag{4.7}$$

$$i_{th} = \left(\frac{4kTB}{R_n}\right)^{\frac{1}{2}} \tag{4.8}$$

$$i_s = (2eI_sB)^{\frac{1}{2}} \tag{4.9}$$

where k is the Boltzmann's constant, B is the detector bandwidth, T is the temperature, R_n is the resistance.

For a simple detector, SNR is defined as,

$$SNR_c = \frac{I_s}{(i_{th}^2 + i_d^2 + i_s^2)^{\frac{1}{2}}} \tag{4.10}$$

Lidar equation is used to estimate the amount of backscattered laser power that reaches the receiver surface. The current I_s produced by the detector, for the calculated return laser power incident on its surface, can be calculated by using the responsivity of the chosen detector.

4.1.8 NOISE EQUIVALENT POWER (NEP)

Noise equivalent power or NEP is the amount of incident optical power that produces a detector output current equal to the noise current of the detector. NEP is a measure of how sensitive our detector to weak return signal. A low value of NEP is required to ensure sufficient sensitivity to detect weak signals from far away (Figure 4.4). It is represented in units of watts per square root hertz. NEP is defined as,

$$NEP = \frac{IA}{SNR * B^{\frac{1}{2}}} \qquad (4.11)$$

where, I is the irradiance incident on the detector in watts, A is the area of the detector in cm² and B is the measurement bandwidth. Ideally, the measurement bandwidth is the same as the detector bandwidth. It can be larger than the detector bandwidth depending on our sensing application. However, the noise contamination is higher for larger bandwidth because most noise sources are wide band sources, i.e., noise power contributions are higher at higher frequencies.

4.2 AVALANCHE PHOTODIODE (APD)

APD are a class of photodetectors [26] that are typically used to detect very weak signals. For lidar applications, a linear mode APD is preferred to help detect signals from targets far away from it. The APD has an internal gain mechanism which attracts users for their specific application. For every photon incident on the detector surface, a series of electrons are produced by an avalanche effect. This is a process where one electron released by absorbing the incident photon leads to a chain reaction, where more electrons are knocked off and an avalanche of electrons are released and driven out as current through the detector circuitry (Reverse Biased). It is commonly referred to as impact ionization in the literature. The detector is said to be working in linear mode when the response is proportional to the input optical irradiance. However, it is possible to drive the APD at a reverse bias voltage greater than its breakdown voltage, when the response is non-linear. Such APD's are commonly referred to as Geiger mode APD's.

When operating in Geiger mode, any false trigger due to a background photon is also amplified by the APD. It does not have the capability to differentiate between the signal photons and the background photons. Responsivity of the detector is a function of wavelength and hence, any source that produces photons at a specific wavelength will trigger an event. Typically, an APD gain is in the order of 20-40. There are Silicon APD's and InGaAs based APD's. For 1550nm as mentioned in the earlier sections, an InGaAs APD is necessary as its responsivity is maximum at this wavelength.

Using the lidar equation the total return power (P_{return}) can be calculated for a given range. The current produced by the receiver APD sensor is given by,

$$i_{signal} = P_{return} \times R_o \times M \qquad (4.12)$$

where M is the APD gain and R_o is the responsivity.

This current signal is then converted to a voltage value using a transimpedance amplifier. The output voltage is given by,

$$V_{out} = i_{signal} \times R_{TIA} \qquad (4.13)$$

where R_{TIA} is the resistance of the transimpedance amplifier.

Internal gain of the APD contributes to the noise in the signal. The major sources of noise in an avalanche detector are signal (true signal + background) shot noise, dark current shot noise, thermal noise and amplifier noise (TIA). The most dominant noise source for an APD is its shot noise. It has two main components, signal shot noise and dark current shot noise, both of which are amplified by the internal gain. Thermal noise is negligible. Also, APD's have an excess noise factor that arises from the statistical nature of the impact ionization process. It is a measure of the noise in an APD.

The noise current is given by [56],

$$i_n = \left(\sqrt{2qI_dM^2F + 2q(P_{total})R_oM^2F} + i_{tia}\right)BW \qquad (4.14)$$

$$F = M\left(1 - (1-k)\left(\frac{M-1}{M}\right)^2\right) \qquad (4.15)$$

where I_d is the dark current (statistical mean value), i_{tia} is the noise current from the external amplifier (TIA), M is the APD gain, F is the excess noise factor, q is the charge of an electron, k is the Boltzmann's constant, R_o is the responsivity, P_{total} is the total return signal power and BW is the bandwidth of the system. APD has its associated bandwidth and TIA has its own bandwidth; the system bandwidth would be the smaller of the two.

The noise-equivalent power for an APD is given by,

$$NEP = \frac{\left(\frac{i_{noise}}{\sqrt{BW}}\right)}{R_oM} \qquad (4.16)$$

The NEP of a detector is the power of the weakest signal that can be detected. This specifies the maximum range of operation for the lidar prototype, since signal strength depends on the range. Figure 4.1 and 4.2 shows the APD current and SNR as a function of range. We observe that as the target goes farther away, the return signal SNR decreases.

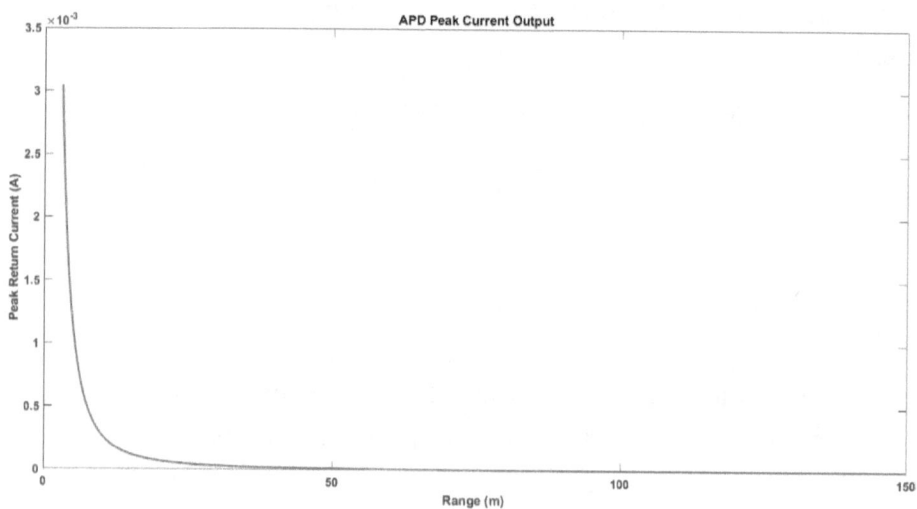

Figure 4.1 APD returns signal peak current (Equation 4.12) Vs range

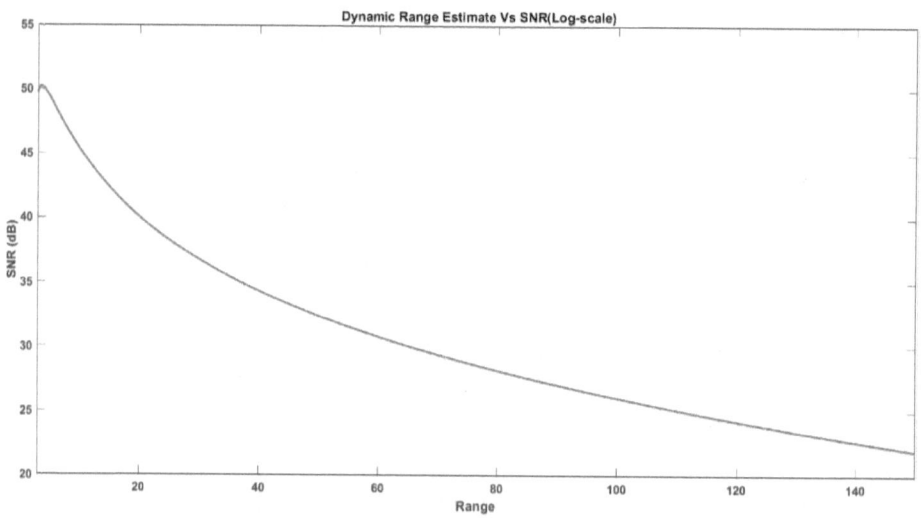

Figure 4.2 Dynamic Range: Signal to noise ratio (dB) current Vs range

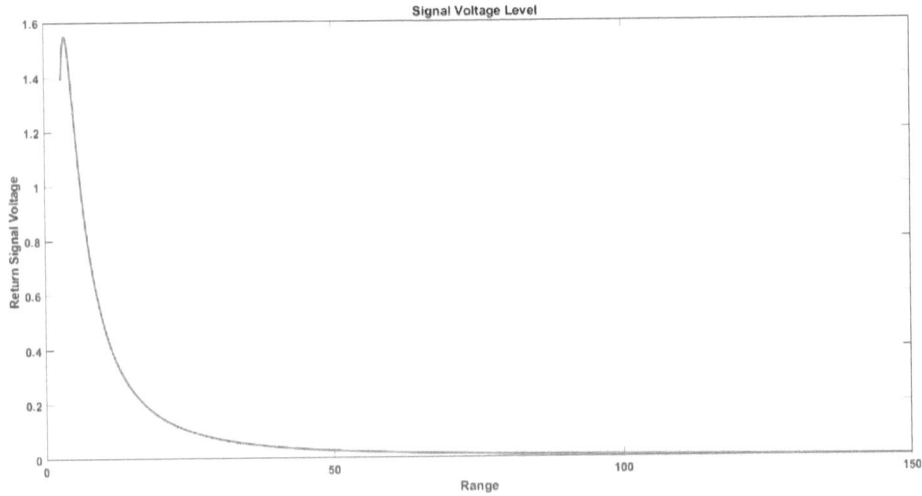

Figure 4.3 Return signal peak voltage (Equation 4.13) Vs range

The parameters used for the above simulation are listed in Table 4.1. An off-the shelf APD-TIA solution was chosen for the lidar prototype.

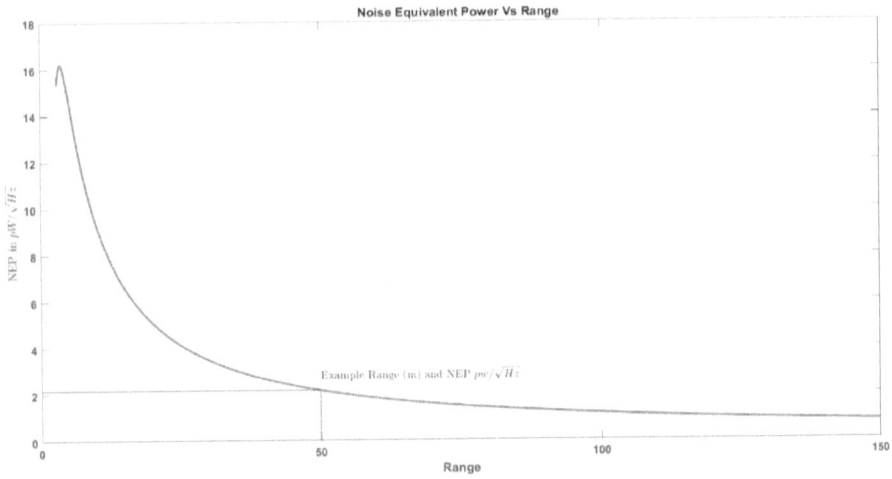

Figure 4.4 Detector Sensitivity – NEP Vs range

Table 4.1 Lidar Simulator Parameter

SIMULATOR PARAMETERS	SPECIFICATION
Peak Power	300
Target Reflectivity	0.1 (10%)
Receiver Lens Diameter	0.0254 m
Extinction Coefficient	0.12/1000 per meter
Optics efficiency (T/R)	70%
Wavelength	1550 nm
APD Gain (M)	20
i_d Dark Current	1 nA
i_{tia} TIA noise Current	25 nA

Table 4.2 Example of Lidar Specifications

LASER SOURCE	Pulse Width $\rightarrow$ 5 − 20 ns Peak Power $\rightarrow$ 20 mW PRF $\rightarrow$ 1 kHz − 1 MHz
LASER AMPLIFIER	Erbium Yttrium Doped Fiber Amplifier (@ 5ns, 750 kHz $\rightarrow$ 381 W *peak power*)
COLLIMATOR	$\theta_{div} \rightarrow 0.016°$ Diameter $\rightarrow$ 7 mm
MEMS mirror	Diameter $\rightarrow$ 4 mm FOV $\rightarrow$ XFOV $(-10°, 10°)$, YFOV $(-10°, 10°)$ Average Power Limit $\sim$ 2W Scan Speed $\rightarrow$ up to 100 rad/sec
APD detector + TIA	Bandwidth $\rightarrow$ APD (10 GHz), TIA(750 MHz) $R_o \rightarrow 0.95$ APD (Gain) $\rightarrow$ M = 10 − 20 TIA(Gain) $\rightarrow$ 1200 Ω

CHAPTER 5: TOF SYSTEM DESIGN: SIGNALS & TIMING

Another major component of a lidar system is a timing device. The goal of the timing device is to accurately measure the Time of Flight (TOF) of the return pulses and calculate the range of a target. There are two typical approaches for time measurement. One is to use a Time to Digital Converter (TDC) to precisely time the signals by a very fast counter to the order of a few tens of pico-seconds. Another approach is to use a high sample rate Analog to Digital Converter (ADC) to capture the shape of a signal and calculate its arrival time. One of the aims of the ADC approach is to utilize shape information to improve the signal detection in higher noise. In this chapter, the fundamentals of a TDC and an ADC, and their corresponding timing algorithms, will be introduced.

5.1 TIME TO DIGITAL CONVERTER

5.1.1 Fundamentals of Time to Digital Converter

A TDC has two key components: a Schmitt comparator and a timing circuit, as illustrated in the block diagram in Figure 5.1. The comparator generates logic signals by comparing the analog signal with a predefined threshold voltage. The timing circuit records the timestamp of the point when the logic signal from the comparator changes value from 'low' to 'high'. This circuit consists of a coarse counter, which tracks the number of clock cycles that have been completed, and a fine counter or ring oscillator, which tracks the number of sub-divisions corresponding to the arrival of the start and stop signal within a single clock cycle. With the use of these two counters, the TDC can obtain a time resolution of picoseconds.

5.1.2 TDC-based timing algorithms

Timing methods for TDCs generally fall into two main categories: the leading-edge timing algorithm and the constant fraction discriminator (CFD).

A. Leading-edge timing algorithm

In leading-edge timing algorithms, the timestamp is recorded when the leading edge of a signal crosses the threshold of the comparator. Since this measurement is carried out at the leading edge of a signal, this time detection method will still be functional even if the signal saturates a photodetector.

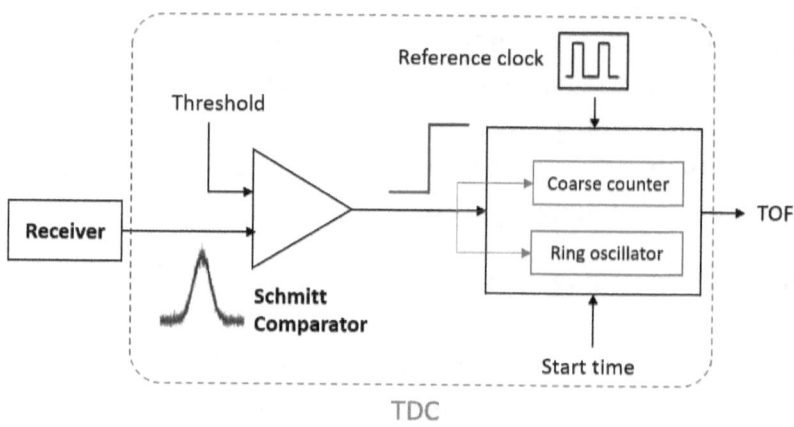

Figure 5.1 Block Diagram of a TDC

Leading-edge detection is not limited by the dynamic range of a photodetector. TDC-based timing techniques are instead limited by other inherent noise and errors. If the signal is noise-free, the time when the threshold is exceeded is constant. However, amplitude variation is always present due to various noises, in which case the measured TOF contains uncertainty. This kind of noise is called 'jitter'. An example of the jitter is shown in Figure 5.2. Mathematically, this noise is defined as [58]:

$$\sigma_{jitter}^2 = \frac{\sigma_{noise}^2}{(dV/dt)^2} = \frac{t_r^2}{SNR^2} \qquad (5.1)$$

Thus, jitter has an inverse relationship with the slew rate (dV/dt) at the trigger position.

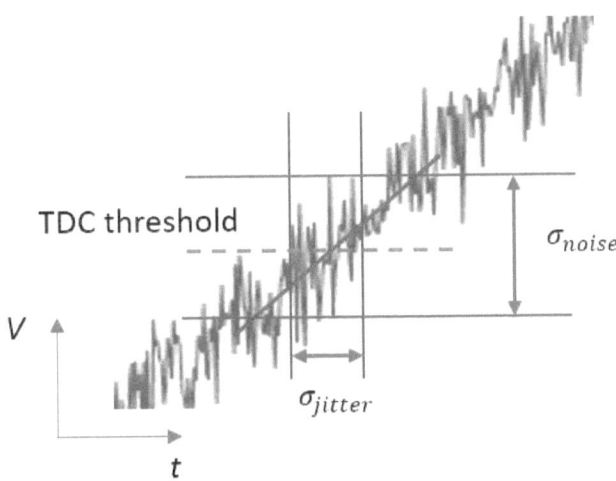

Figure 5.2 Schematic of Jitter Noise. The black curve is a signal with noise. (Figure adapted from [59]).

Another type of error the leading-edge timing method may have is 'walk error' which is also due to variation in the signal amplitude. As illustrated in Figure 5.3, problems arise when two signals are reflected by targets at the same distance but with different return powers. If the comparator threshold and rise-time of the two signals are kept constant, the signal with a smaller amplitude triggers the threshold later than the larger one, which leads to a difference in the arrival time between the signals. Walk error refers to this time difference due to the difference in amplitudes and is another major contributor to timing inaccuracy of leading-edge timing algorithms. Various methods can be used to compensate for the walk error, including 'gain control compensation' [61], 'time-over-threshold compensation' [62], 'slew-rate compensation' [63], and the 'time-variant threshold' method [64].

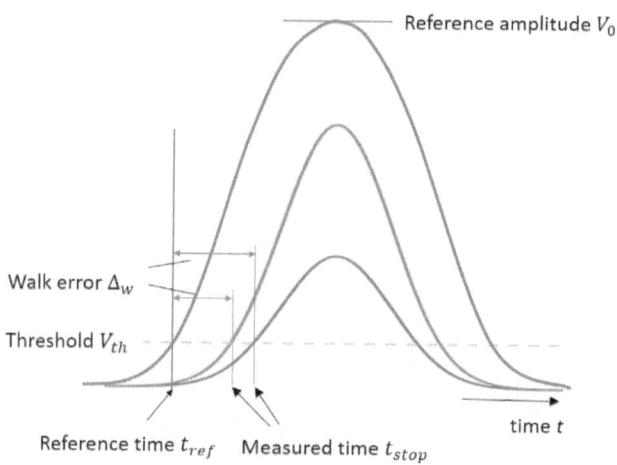

Reference amplitude V_0

Walk error Δ_w

Threshold V_{th}

Reference time t_{ref} Measured time t_{stop}

time t

Figure 5.3 Illustration of a Walk Error

B. Constant Fraction Discriminator

The Constant Fraction Discriminator (CFD) is another timing algorithm for TDCs. This method is often used as it can help to reduce jitter noise and walk error. As shown in Figure 5.4, traditional CFDs copy the input signal into two otherwise identical signals, attenuating and inverting one while delaying the other. The attenuated signal is attenuated by a constant factor corresponding to the desired fraction of the full amplitude. This value is usually located where the original signal has the largest slope, often between 20% and 40%. The time-delayed signal is tuned to ensure the fraction point on the leading edge of the delayed signal aligns with the peak of the attenuated signal. Subsequently, the two delayed and attenuated signals are added to generate a bipolar signal, which is then passed to a zero-crossing comparator to generate the logic signal for time discrimination. The zero-crossing point corresponds to the timestamp at the optimal fraction point on the delayed signal, which is also the time at the peak of the original signal. In addition, a leading-edge arming discriminator triggered above the noise floor of the signal is applied to prevent the zero-crossing comparator from being mistakenly triggered on the noise floor preceding the zero-crossing point.

74

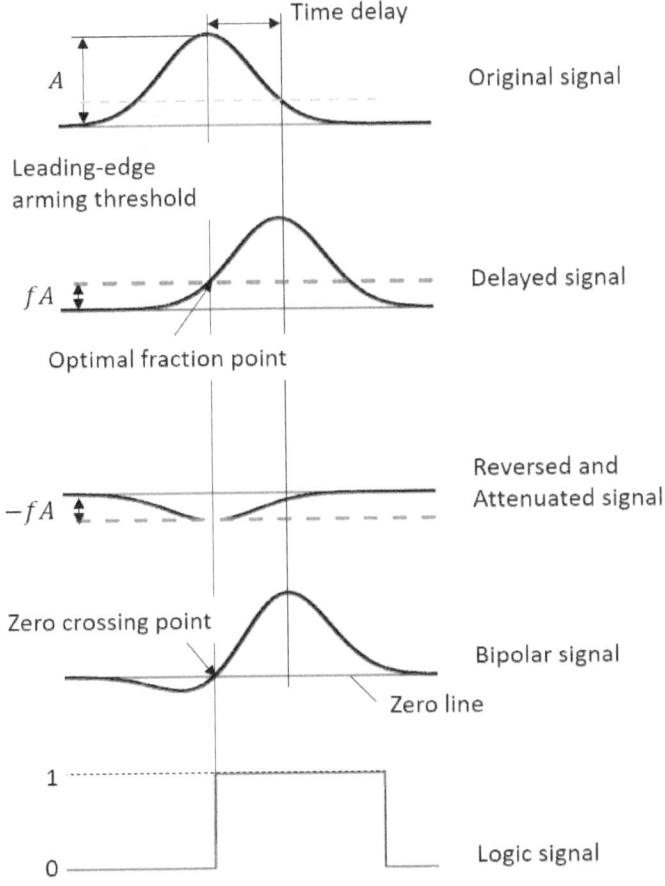

Figure 5.4 Schematic of the Principle of a Traditional CFD.

The major advantage of the CFD is that it theoretically eliminates the walk error due to the independence of the zero-crossing point on signal amplitude as shown in Figure 5.5. However, this approach falls short when an incoming signal saturates the detector or when multiple returns come back in a minuscule time difference. In these cases, the top of the pulse is flattened which can result in a large walk error in TOF measurements.

In addition to reducing walk errors, CFDs also result in decreased levels of jitter noise due to multiple reasons. First, the trigger point is set optimally at the sharpest position of a signal, which reduces the variance of jitter noise. Second, because of its statistical nature, the noise on the attenuated and the delayed signals cancel each other in the summation process. Moreover, since only the leading edge and the peak of the signal are involved in the summation, the CFD is not affected by the pulse symmetricity.

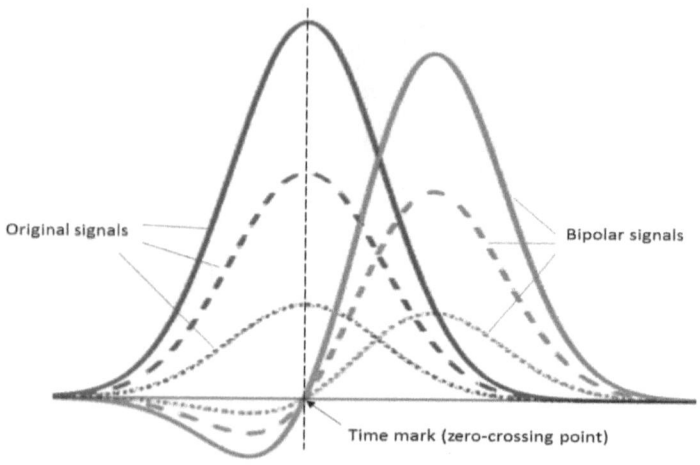

Figure 5.5 Independence of the Zero-Crossing Point on Signal Amplitude.

5.1.3 Example: Texas Instruments TDC 7201

In this section, the TDC 7201 from Texas Instruments (TI) will be used as an example to illustrate basic principles. This TDC was used in our group to practice measuring time of flight. This section uses References in [9].

A. Operational Features

The Texas instruments TDC-7201 is an off-the-shelf timing solution that can be controlled using a microcontroller, such as the TI MSP 430. The TDC solution comes with a graphical user interface that is straightforward to run on any standard windows operating system. This specific TDC solution has two independent TDC's in the integrated circuit. Each TDCx (x = 1, 2) has separate start and stop channels. The start signal is fed into the START port on the TDC-x and the return signal from the avalanche photon-detector is connected to the STOP port of the TDCx. A picture of the TDC-7201 is given in Figure 5.6.

Figure 5.6 TDC7201 EVM – Evaluation Board from Texas Instruments [9]

The TDC board is powered by the microcontroller board. The microcontroller settings and the TDC settings can be set by the user. The performance of a TDC is gauged based on its timing resolution. The timing resolution for the TDC 7201 is rated at 55 ps, i.e., the TDC can provide a range resolution to the order of few millimeters. The TDC unit uses a clock rated at 8 MHz. Thus, the clock period is 125 ps. TDC undergoes an internal calibration after a user-specified time duration (typically 10 measurements) to address clock drift and jitter over time.

Each TDC has two modes of operation: Mode-1 (Short Range) and Mode-2 (Long Range). The Mode-1 setting is used to measure the time of flight between 12 ns to 2000 ns. This translates as 1.8 m to 300 m in range. The Mode-2 setting is used to measure the time of flight values between 250 ns to 8 ms, i.e., 37.5 m to 120×10^4 m. Another additional mode that is available with this TDC uses both the TDC's (x = 1, 2) simultaneously to measure the time between 0.25 ns to 8 ms. This is referred to as the combined mode of operation.

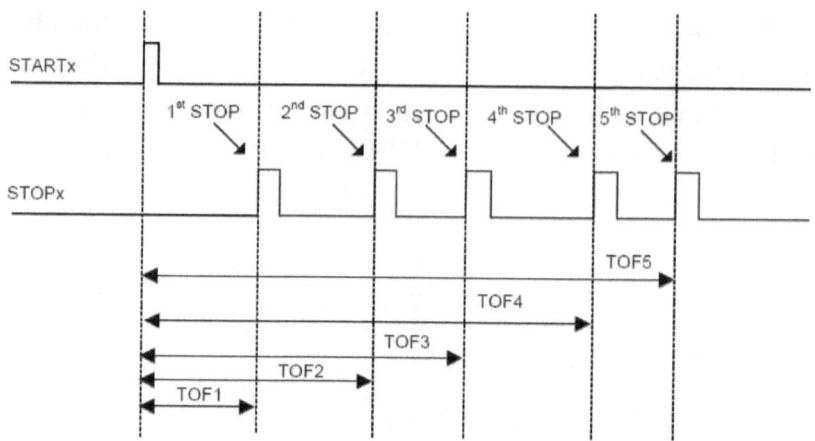

Figure 5.7 TDC Measurement Technique – Basics, STARTx and STOPx (x = 1, 2) are the start and stop signals for TDCx (x = 1, 2). See Ref [9].

The START and the STOP signal are input into the TDC through a standard 50 Ω SMA female connector on the TDC board. These analog signals are connected to one of the leads of a Schmitt comparator. The other lead of the comparator is connected to a reference/threshold voltage. The incoming signal (START or STOP) voltage is compared with the threshold to produce a mixed signal - an analog signal with digital-like values. Specifically, a mixed signal has a high value (i.e., bit → 1) when the incoming signal is greater than the threshold and a low value (i.e., bit → 0) when the incoming signal is lower than the threshold. At the exact moment when the incoming signal crosses the threshold, the coarse counter and the ring oscillator counts are captured and stored in the local memory registers. These values are retrieved for TOF computation between the START and the STOP signals.

TDC also specifies requirements on the rise time and the pulse width for both the START and the STOP signals. The two input signals require their rise time to be within 1 ns – 100 ns. The fall time of the two signals is also required to be within 1 ns – 100 ns range. Similarly, the pulse width requirements for both the START and STOP signals are set at 10 ns. The comparator reference voltage is preset at 1 V. Thus, the

START and STOP signal voltage peak values must be greater than the threshold to start making measurements. The maximum input signal voltage is restricted to 3.3 V to prevent any damage to the boards. These conditions discussed above should be ensured for the TDC to meet its accuracy and uncertainty metrics.

5.2 ANALOG TO DIGITAL CONVERTER

5.2.1 Fundamentals of Analog to Digital Converter

A. Digital Processing

An ADC-based time device uses an ADC to convert analog signals from a photon-detector to digital signals, determine if the signal is a return pulse or noise, and calculate the TOF if it is a pulse. The structure of the ADC-based timing method is shown in Figure 5.8. The digitization process of an ADC consists of two steps: sampling and quantization. The first step is to sample an analog signal in time by a sample-and-hold device that takes and holds a value from the continuous signal every T_s seconds. The time interval T_s is called the sampling interval, and the sampling rate is denoted as $f_s = 1/T_s$. The resulting signal has a discrete time sequence but still has continuous values of amplitude. In the quantization step, the continuous values of the amplitude are converted to discrete values by a quantizer. Therefore, a signal with discrete values in both time and amplitude is generated, and such signals are referred to as 'digital'.

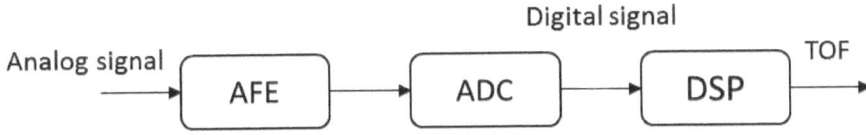

Figure 5.8 Schematic of ADC-Based Timing method

B. Quantization and Sampling

In the quantization process, an important parameter of the quantizer is the resolution, which indicates the number of discrete levels

79

it can produce over the dynamic range of the signal amplitude. The resolution is usually expressed in bits. For example, for a 6-bit ADC, there are 2^6 discrete levels to store the signal amplitude. The difference between two successive quantization levels is called quantization steps [65].

$$\Delta = V_{k+1} - V_k \qquad (5.2)$$

Thus, the quantization steps Δ and the ADC resolution B are related to the dynamic range V_{dy} of the input signal by

$$V_{dy} = 2^B \qquad (5.3)$$

The nature of the quantization process is to approximate an analog signal by discrete values. The finite number of quantization levels introduces error between the digital signal and the analog signal, which is denoted as the quantization error.

The sampling rate of an ADC affects the reconstruction of an analog signal. As known from the Nyquist's theorem, the sampling rate should be equal to or larger than twice the largest frequency (Nyquist frequency) of the signal to correctly represent the signal, i.e., $f_s \geq 2f_{max}$. Otherwise, aliasing occurs. However, the bandwidth of the input signal of an ADC may not always be known precisely, so the Nyquist criterion may not be met in practice. Therefore, to prevent aliasing, a low-pass filter is usually placed before an ADC, to truncate the frequency components larger than the Nyquist frequency. Such a filter is called an anti-aliasing filter. Alternatively, oversampling of the signal to a rate much higher than the Nyquist frequency is also often used to reduce the aliasing error.

5.2.2 Timing Algorithms

The timing algorithms using an ADC can be achieved by applying digital versions of the TDC-based timing algorithms to digital signals, or by taking advantages of the shape information contained in the digital signals.

A. Digital Version of TDC-Based Timing Techniques

The digital versions of TDC-based timing algorithms include all the algorithms mentioned above like leading-edge detection techniques and CFD's. The difference from the TDC-based algorithm is that the algorithms take digital signals. The details of those methods are described in Section 5.1. A drawback of the digital algorithms is that since the digital signals only have discrete values, the TDC trigger threshold may not always coincide with the sample points of a digital signal. In other words, the exact trigger point may fall in between two successive samples. In this case, interpolation between neighborhood samples is needed to find the exact time mark corresponding to the threshold value. Linear interpolation is a common selection, but it could also affect the timing accuracy, depending on the local linearity of the signal [65].

B. Detector and Estimator

The digital version of TDC-based timing methods utilizes the ADC as a TDC. This approach limits the potential of an ADC because the shape information contained in the digital signal is valuable but ignored in TDC methods. Alternatively, ADC-based timing algorithms can take advantage of the signal characteristics. ADC-based algorithms are generally composed of a detector and an estimator. The detector takes digital signals from an ADC as an input and determines the presence or absence of a transmit pulse from the signal contaminated by noise.

Generally, we define a hypothesis $\mathcal{H}_1$ if the signal of interest (a return pulse in our case) is detected, and a hypothesis $\mathcal{H}_0$ if the input signal is noise. If a pulse is detected, an estimator is applied to estimate the arrival time of the return pulse. There are different ways of utilizing the shape information in the ADC-based timing algorithms, for example,

81

using the pulse width or the centroid of the signal and the Neyman-Pearson (NP) detector which is also called the Matched Filter.

C. Neyman-Pearson (NP) Detector

The NP detector was first introduced by Jerzy Neyman and Egon Pearson, and has shown to be the favorable detector for radar / laser signal detection [66, 67, 68]. Kay [68] has provided a detailed explanation of the NP detector, therefore we provide only a brief description here. In general, the NP detector follows the Neyman-Pearson theorem that maximizes the probability of detection of a return pulse subject to a required probability of false alarm (PFA). The probability of false alarm is defined as the percentage of false-positive signals to the total number of noise signals. The NP detector conducts a likelihood ratio test to evaluate if the likelihood of a test signal being $\mathcal{H}_1$ versus being $\mathcal{H}_0$ is larger than a threshold defined by the required PFA. If the likelihood ratio is larger than the threshold, the test signal is a return pulse. Otherwise, the test signal is noise. Mathematically,

$$\text{if } L(\vec{x}) = \frac{p(\vec{x}; \mathcal{H}_1)}{p(\vec{x}; \mathcal{H}_0)} \geq \gamma \text{: the signal is } \mathcal{H}_1; \tag{5.4}$$

$$\text{if } L(\vec{x}) = \frac{p(\vec{x}; \mathcal{H}_1)}{p(\vec{x}; \mathcal{H}_0)} < \gamma \text{: the signal is } \mathcal{H}_0, \tag{5.5}$$

where $L(\vec{x})$ is the likelihood ratio and γ is the threshold. The test signal $\vec{x}$ has N sample points, i.e., $\vec{x} = \{x[0], x[1], \dots, x[n]\}$ $(0 \leq n < N)$ and $p(\vec{x}; \mathcal{H})$ is the probability of the signal being $\mathcal{H}_1$ or $\mathcal{H}_0$. Assuming the noise on the signal is Gaussian distributed white noise, the above equation can reduce to

$$\sum_{n=0}^{N-1} x[n] \, s[n] > \gamma'. \tag{5.6}$$

The template of the signal of interest is denoted by $\vec{s} = \{s[0], s[1], \dots, s[n]\}$, and the threshold γ' can be derived from the pre-defined PFA and the noise variance. This equation presents how an NP detector works in general, that is, the NP detector computes a cross-correlation of the test signal with the pulse template. Then, the correlation is compared to a threshold to determine the presence of a return pulse. If a pulse is detected, the timestamp of the sample at which the correlation reaches its maximum is the arrival time of the return pulse [67].

5.2.3 Example: Cronologic 5G ADC

In this section, the Ndigo5G ADC from Cronologic will be used as an example to illustrate features and configurations of a specific ADC available on the market today.

A. Operational Features

The important functions an ADC may have include: 'channel modes', 'analog offset', 'zero-suppression', and 'trigger modes'. An ADC usually has multiple channels taking analog inputs, and the channels could share one or multiple clocks. The 'channel mode' is the combination of channels that can be used. Taking a 4-channel ADC (Channel A, B, C, D) with a clock of 5 GHz as an example, the ADC can be operated in 1-channel mode (Channel A, B, C, D), 2-channel mode (e.g. AC, BD) and 4-channel mode (e.g. AAAA, ABCD). If all the channels share one clock, the 4-channel mode has a sampling frequency of 1.25 GHz signal, one-fourth of the speed of 1-channel mode (5 GHz). In the application of the TOF estimation, the 2-channel mode is commonly used which takes an analog signal from a laser source (START signal) and a return signal from a photon-detector (STOP signal). Another feature of an ADC is the analog offset. Using the analog offset function, users can remove the DC component of the signal to keep the signal remain within the ADC range, which maximizes the dynamic range. The zero-suppression function is important for an ADC to reduce the load of data transfer. The zero suppression means that an ADC only starts to acquire data when the predefined ADC threshold is exceeded, and the stored waveform data is referred to as 'packet'. A packet contains the sampled data and the absolute time of the last sample of the packet. The absolute time is measured since the data acquisition starts. The timestamp of each sample of a signal can be calculated using the time interval between the successive samples and the number of samples.

An ADC could have different trigger modes to collect data and two examples are edge-trigger mode and level trigger mode. The edge and level trigger modes are illustrated in Figure 5.9. The edge-trigger mode specifies the number of samples after the trigger point, which is denoted as the 'collection length'. All the data collected using the edge-trigger mode has the same length, but not all the samples are guaranteed to be above the threshold. On the contrary, the level-trigger mode saves all the samples above the threshold instead of saving a fixed number of samples. It means the length of the stored data is dependent on the signal amplitude. Both modes can also save some samples before the ADC trigger point, which is called 'precursor'. In addition, the level-trigger mode can also save samples after the sample becomes lower than the threshold, and the number of such samples is called 'post-cursor'.

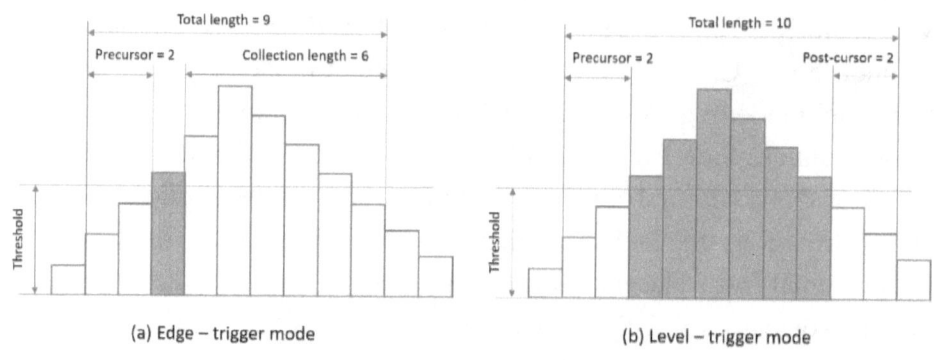

(a) Edge – trigger mode   (b) Level – trigger mode

Figure 5.9 ADC Trigger Modes (Adapted from [69]).

5.3 COMPARISON: TDC VS ADC TIMING METHODS

TDC-based approaches have significant advantages over the ADC-approaches, primarily simplicity and cost. Due to this simplicity, they are generally fast, and consequently are not a bottleneck for processing large numbers of pulses in real-time. Without the need to store a large number of samples or to employ complicated DSP circuits, TDC's require less power, and typically cost much less than ADC's. Additionally, TDC-based algorithms are less affected by the saturation of a photon-detector, and fast timing circuits also allow a much higher time resolution than what an ADC can achieve without incurring much higher costs. However, TDC-based algorithms also require compensation methods to reduce walk error. Simple TDC algorithms are unable to distinguish noise from true signals when the noise amplitude exceeds the threshold.

ADC methods, on the other hand, provide much greater flexibility to utilize detailed pulse shape information, potentially allowing better detection in high noise conditions. However, dynamic range adjustments would be required for ADC algorithms over practical ranges for autonomous vehicles. This is the case because as pulse returns decrease in size with greater ranges (according to the lidar equation), the ADC needs to relocate its available voltage levels lower to achieve similar pulse shape resolution. The adjustment is typically achieved using an RC circuit which is designed to decay in a similar manner as pulse power reduces with the time of its return.

CHAPTER 6: TOF SYSTEM DESIGN: BEAM STEERING

A lidar system must be able to successfully map the area around a self-driving car or the interiors of a room accurately and quickly. To do so, it must be able to map multiple points within the field of view of the lidar sensor. Thus, in order to generate a 3D point cloud of ranges, the time of flight system requires a scanning module. There are presently a multitude of approaches for quick and efficient scanning of the laser beams in commercially available sensors.

6.1 MECHANICAL SCANNING

Mechanical scanning lidar systems use rotating mirrors to steer the laser beam or alternatively physically rotate the entire sensor (both source and receiver) as in large-scale galvanometer-based optical scanners. This creates the potential for a wide field of view and due to the use of mechanical components also yields a high signal to noise ratio. For example, the Velodyne HDL-64E sensor [57] has 64 laser diodes stacked on a rotating platform that rotates the vertical stack of laser beam around an entire 360° FOV. However, these systems are often prohibitively expensive (on the order of thousands of dollars for the rotating assembly and associated highly-focused optics) as well as bulky and complex. This makes mechanical scanning risky in terms of production and failure in harsher environments [70]. Thus, current commercial applications of lidar to the automotive industry are increasingly moving towards a solid-state approach to beam steering.

6.2 SOLID STATE SCANNING

Solid state lidar systems direct the laser beam to various positions throughout the field of view without mechanically moving the source or any macroscopic optics. This lack of moving parts makes solid state approaches significantly more straightforward, reliable, and cost-efficient than the mechanical approach. However, this setup does result in a limited field of view, which necessitates the synthesis of multiple individual sensors to create a full 360 field of view as is required for vehicular applications. Presently, there are several mechanisms for implementation of solid-state beam steering, including MEMS mirrors, optical phased arrays, and flash lidar approaches.

6.2.1 MEMS MIRROR SOLUTIONS

In a lidar system deploying MEMS mirrors as the beam steering solution, the tilt angle of microscopic mirrors is varied rapidly and driven electromechanically. Due to the mirrors' extremely small size, they exhibit both low moments of inertia and high resonant frequencies, which results in smaller driving power requirements, improved robustness to environmental conditions, and lower cost of production. These systems can scan an entire field of view in fractions of seconds and easily adjust the resolution of their scan patterns to focus on a specific area within the field of view. One concern with this approach however lies in the decreased diameter of the scanning mirror, which acts as a small aperture resulting in increased diffraction and thus divergence of the beam and ultimately limiting the resolution of the overall system.

6.2.2 OPTICAL PHASE ARRAY SYSTEMS

The optical phase array (OPA) system steers the laser beam to various positions in the field of view with an array of optical antennas. These antennas are illuminated with coherent laser light and re-emit it while individually altering the phased and amplitude, for example creating a constant phase shift between each array element. This allows for control of the optical wave front shape including the positions of the far field maxima and minima of interference, thus choosing the direction of the resulting beam. Altering the phase shift changes these positions and steers the beam. A similar technology has been used for decades with phased array radar systems with many applications including military, acoustics, and astronomy.

However presently the range of these OPA systems is severely restricted due to high component losses which restricts power.

Nonetheless, while the OPA technology is still in its nascent stages, it is actively growing within the research and development communities both at university and industry level. For example, the startup company Quanergy based in Sunnyvale is reportedly developing a phased-array lidar solution [71].

6.2.3 FLASH LIDAR SYSTEMS

Flash lidar systems do not require a scanning module as have the previously discussed beam steering solutions. Instead, these sensors operate on a similar principle to digital flash cameras, with sensing elements that measure time-of-flight instead of light intensity. A single pulse of an expanded, large area laser beam illuminates the detector field of view. This specialized detector is located in the focal plane of the system and consists of an array of avalanche photodiodes to detect the backscattered light. Instead of a point-by-point scanning of the field of view, these systems gather time of flight data points of an entire horizontal plane simultaneously with each flash. Thus, flash lidar systems are less vulnerable to timing distortion issues associated with small movements or vibrations of the sensor or its surrounding environment. They are also quite fast with a high data capture rate.

Nonetheless, the flash approach has a few serious drawbacks. In order for the entire scene to be illuminated and for the system range to be far enough, an extremely high peak laser power is required. This currently restricts the effective range of flash lidar to a few tens of meters. In addition, the resolution of this type of system is limited by the number of receivers in the detector array. If the total number of receivers on a given array is increased in an attempt to increase resolution, their size must be reduced which also decreases the number of photons that receiver capture, resulting in a lower signal to noise ratio. This tradeoff limits overall resolution and must be carefully balanced in flash systems [71].

6.3 LIDAR PROTOTYPE SYSTEM MEMS SOLUTION

The role of the lidar prototype designed here is to work in tandem with a camera sub-system to identify and classify regions within the FOV with more certainty. Thus, our requirement demands that the scanning modality is quick and highly repeatable to provide data throughput at a higher frame rate. A full 360degree FOV is not required for this application, but the potential for dynamic resolution control is quite useful. Also, the scanning solution should to be compact, easily integrated with the current prototype, and not cost-prohibitive. Based on these requirements the best solution available within a short lead time and off-shelf is a MEMS (Micro-Electro Mechanical System) mirror.

Specifically, in the MEMS solution used with this prototype, the user specifies a scan pattern from which a series of digital voltage representations are calculated. These digital voltages are then run through a digital-to-analog converter, resulting in a set of analog drive voltages which dictate the mirror tilt, steering the beam in the particular scan pattern.

6.3.1 RESONANT MODES

This MEMS solution is gimbal-less and dual-axis (tip-tilt). It has three possible modes of operation. In the quasi-static, point-to-point mode, each steady-state, analog driver voltage maps to a steady-state angle of rotation of the mirror resulting in a highly repeatable pattern with the possibility for constant scan speeds. This mode requires more complex control algorithms and limited scan angle and speed in comparison with the dynamic mode of operation [74].

In the dynamic, resonant mode, control design is simplified and larger scan angles can be achieved at high frequencies. The micromirror is supported by single-crystal silicon springs which also serve as the restoring force during actuation, forming an entire second-order spring-mass system. Low driver voltages are modulated at frequencies close to the resonant device's resonance (usually on the order of kHz but dependent on various micromirror parameters including diameter) in order to produce these larger beam angles, up to -16° to 16° [72]. Use of this mode produces sinusoidal scan trajectories with nonuniform speed and can damage the device if improper settings are used such as exceeding maximum allowable bias voltage levels. Since our micromirror solution was designed primarily for point-to-point mode, when driven near resonance it risks exceeding safe operating angles and thus near or at resonance operation, significantly lower voltages and with additional care is used in operation. Scan patterns in this mode are Lissajous motions such as circles, ellipses, and higher-order patterns which can be adjusted to cover the imaging area.

With our MEMS solution, there is also a third mixed mode of operation in which one axis is in quasi-static mode and the other is in resonant mode, which allows for raster scan patterns of the specified imaging area. To create this raster pattern, the resonant axis is run quickly creating horizontal lines, while the point-by-point axis is run in a slower sawtooth pattern [72]. As in resonant mode, careful examination of parameters for the resonant axis is vital to prevent damage to the device.

In the figure below [72], each of the three scanning modes can be observed. The first image is a point-to-point scan. The second is a mixed mode scan in which the x-axis motion is a sinusoidal wave in resonant mode and the y-axis motion is a triangular wave in point-to-point mode. The third is a resonant Lissajous scan pattern generated by both axes run in resonant mode.

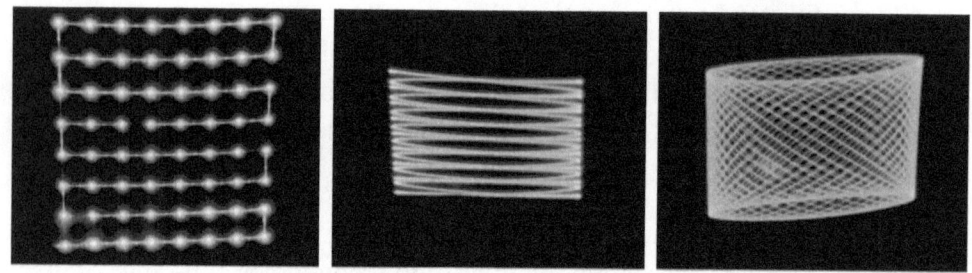

Figure 6.1 MEMS scanning patterns.

6.3.2 ARRAY ALIGNMENT AND SYNCHRONIZATION

Clearly the advantages produced by the resonant scanning modes of the MEMS mirror are interesting for autonomous vehicle applications. Greater scan angles provide a larger field of view, high speed scans allow for high data capture rate, and the low power requirements for large scan areas are favorable for system design. However, in our prototype we have encountered some serious stumbling blocks associated with this approach primarily due to the nonuniform velocities of the resonant scan patterns.

In order to obtain a 3D point cloud, the range values obtained from the timing of the detector's capture of the backscattered light must be aligned with direction in which the laser beam is pointing. However, this alignment is far from straightforward. The micromirror is provided with analog drive voltages via a USB controller which interfaces with the user software to take scan pattern specifications and store waveform patterns in a buffer which are then used to drive the mirror [73]. This buffer creates an undefined amount of lag in the mirror positions, allowing for relative but not absolute knowledge of the laser beam's direction as a function of time. Thus, an array of mirror positions ordered correctly relative to each other but with no absolute time stamp is produced in addition to the array of range data points from the detector and the time-digital converter.

There are two possible techniques for approaching this issue. First, a more complete, top-down solution would be to determine absolute time stamps for each element of both the mirror position array and the time of flight array. This would require complete characterization of the lags and nonuniformities associated with the USB controller in every aspect of its communication with and control of the micromirror as well as its manipulation and storage of waveform scan patterns into analog drive voltages. Since the resonant frequencies involve significantly nonuniform scan velocities, these characterizations would be far more complex than a single constant delay factor.

The approach that has produced more preliminary success with this synchronization in our design involves a more bottom-up approach to alignment of the two datasets. A relatively straightforward algorithm can be written to step through various alignments of the two arrays, testing out different delay factors for each individual scan. If the alignment is performed correctly, the resulting point cloud will produce a coherent image of the field of view; otherwise, this image will be jumbled. The correct delay factor can then be chosen based on the resulting image. The previous correct delay can be used to refine the process for the next scan, however this factor will still vary from scan-to-scan. This process requires a substantial amount of post-processing which is not ideal for future applications.

Another concept for this synchronization issue which should be explored is the idea of an experimentally-derived characterization of the lag performed at regular intervals, perhaps once per scan. Placing a small reflective element at known, close range in the field of view would create a datapoint in the time of flight array for which the range is going to be at a minimum. Searching for this minimum and lining it up with the known position of the element in the mirror's position array would automatically align the rest of the datapoints, since both arrays are correctly positioned relative to themselves. However, this would entail a more involved design and placement of an extremely small reflective element, the details of which are relatively complex.

6.3.3 PROTOTYPE MEMS SOLUTION SPECIFICATIONS

The mirror diameter is 4mm and can handle up to 2W of average power. In general, the scan speed of the mirror is inversely proportional to the size of the mirror, i.e. a larger mirror, due to its larger inertia will have a slower scan speed. Also, a larger mirror has a reduced tilt angle and thus a narrow FOV. This makes the mirror size a critical parameter to ensure sufficient FOV and scan velocity. Thus, this mirror size was chosen to be as large as possible to minimize the diffraction-limiting effect while not compromising FOV or scan velocity. The figure below shows the MEMS solution used in our laboratory tests:

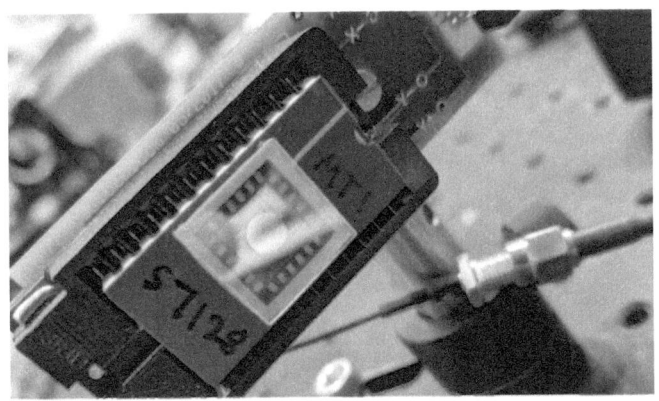

Figure 6.2 MEMS mirror.

The mirror has an anti-reflection coating for the specific wavelength (1550 nm) and ensures very high reflection percentage (greater than 90%). The entire unit is on a compact printed circuit board. A maximum of 130 V is applied to obtain a tilt of 5.5°. Any incident mirror is reflected by twice the tilt angle of the mirror. Thus, the FOV of the MEMS mirror is 20° (-10° to 10°) both along the horizontal and the vertical. The unit provides a 14 bit positional precision, i.e., a tilt resolution of 0.6 millidegrees (10 μrad). The repeatability of the mirror is better than 0.001° at room temperature.

REFERENCES

[1] Thrun, Sebastian, et al. "Stanley: The robot that won the DARPA Grand Challenge." *Journal of field Robotics* 23.9 (2006): 661-692

[2] Frost & Sullivan, "LiDAR: driving the future of autonomous navigation –analysis of LiDAR technology for advanced safety," p. 31, *Frost & Sullivan* (9ᵗʰ February 2016)

[3] Luettel, Thorsten, Michael Himmelsbach, and Hans-Joachim Wuensche. "Autonomous Ground Vehicles-Concepts and a Path to the Future." *Proceedings of the IEEE* 100.Centennial-Issue (2012): 1831-1839.

[4] Bissonnette, Luc R. "Imaging through fog and rain." *Optical Engineering* 31.5 (1992)

[5] Williams, George M. "Optimization of eyesafe avalanche photodiode lidar for automobile safety and autonomous navigation systems." *Optical Engineering* 56.3 (2017): 031224.

[6] McManamon, Paul. "Review of Ladar: a historic, yet emerging, sensor technology with rich phenomenology." *Optical Engineering* 51.6 (2012): 060901.

[7] Amann, Markus-Christian, et al. "Laser ranging: a critical review of unusual techniques for distance measurement." *Optical engineering* 40.1 (2001): 10-20.

[8] McManamon, Paul F. *Field Guide to Lidar*. SPIE, 2015.

[9] TDC7201 Time-to-Digital Converter for Time-of-Flight Applications in LIDAR, Range Finders, and ADAS. *Texas Instruments* [PDF], and Application Report, V. Viswanathan, Short Time Measurement Using TDC7201, SNAA292–June 2016.

[10] Nordin, Daniel. *Optical frequency modulated continuous wave (FMCW) range and velocity measurements*. Diss. Luleå tekniska universitet, 2004

[11] Poulton, Christopher V., et al. "Frequency-modulated continuous-wave LIDAR module in silicon photonics." *Optical Fiber Communications Conference and Exhibition (OFC), 2016*. IEEE, 2016.

[12] Pierrottet, Diego, et al. "Linear FMCW laser radar for precision range and vector velocity measurements." *MRS Online Proceedings Library Archive* 1076 (2008)

[13] Poulton, Christopher V., et al. "Coherent solid-state LIDAR with silicon photonic optical phased arrays." *Optics letters* 42.20 (2017): 4091-4094.

[14] Lum, Daniel J., Samuel H. Knarr, and John C. Howell. "Frequency-modulated continuous-wave LiDAR compressive depth-mapping." *Optics Express* 26.12 (2018): 15420-15435

[15] Dieckmann, Andreas, and Markus-Christian Amann. "Frequency-modulated continuous-wave (FMCW) lidar with tunable twin-guide laser diode." *Automated 3D and 2D Vision*. Vol. 2249. International Society for Optics and Photonics, 1994

[16] Baghmisheh, Behnam Behroozpour. *Chip-Scale Lidar*. Technical Report UCB/EECS-2017-4, 2017.

[17] R. Paschotta, article on 'coherence length' in the *Encyclopedia of Laser Physics and Technology*, 1. edition October 2008, Wiley-VCH, ISBN 978-3-527-40828-3.

[18] Klugmann, Dirk. "FMCW radar in the digital age." *Geoscience and Remote Sensing Symposium (IGARSS), 2016 IEEE Intl*. IEEE, 2016.

[19] Godbaz, John Peter. *Ameliorating systematic errors in full-field AMCW lidar*. Diss. University of Waikato, 2012.

[20] Godbaz, John P., Michael J. Cree, and Adrian A. Dorrington. "Understanding and ameliorating non-linear phase and amplitude responses in AMCW lidar." *Remote Sensing* 4.1 (2011): 21-42.

[21] Laukkanen, Matti. "Performance Evaluation of Time-of-Flight Depth Cameras", 2015.

[22] Adams, Martin D. "Lidar Performance and Calibration Measures for Environmental Mapping."

[23] Whyte, Refael, et al. "Application of lidar techniques to time-of-flight range imaging." *Applied optics* 54.33 (2015): 9654-9664.

[24] Perez, Sara, E. Garcia, and Horacio Lamela. "AMCW laser rangefinder for machine vision using two modulation frequencies for wide measurement range and high resolution." *Testing, Packaging, Reliability, and Applications of Semiconductor Lasers IV*. Vol. 3626. International Society for Optics and Photonics, 1999

[25] Adams, Martin D. "Lidar design, use, and calibration concepts for correct environmental detection." *IEEE Transactions on Robotics and Automation* 16.6 (2000): 753-761

[26] Kasunic J. Keith. *Laser Systems Engineering*. SPIE PRESS. 2016. Print.

[27] Wikipedia contributors. "Transmittance." *Wikipedia, The Free Encyclopedia*. Wikipedia, The Free Encyclopedia, 11 Jun. 2017. Web. 17 Oct. 2018.

[28] Standard, A. N. S. I. "Z136. 1. American national standard for the safe use of lasers. American National Standards Institute." *Inc., New York* (1993).

[29] Wojtanowski, J., et al. "Comparison of 905 nm and 1550 nm semiconductor laser rangefinders' performance deterioration due to adverse environmental conditions." *Opto-Electronics Review* 22.3 (2014): 183-190

[30] Villeneuve, Alain, Joseph G. Lachapelle, and Jason M. Eichenholz. "Pulsed Laser for Lidar System." U.S. Patent Application No. 15/804,997.

[31] Alex Davies "GM Buys a Lidar Startup that could deliver its self-driving future". *WIRED*, 10/09/2017.

[32] Alex Davies "A New Doppler Lidar solves the Self-Driving Car's Need for Speed". *WIRED*.9/05/2018.

[33] Wikipedia contributors. "Optical fiber connector." *Wikipedia, The Free Encyclopedia*. Wikipedia, The Free Encyclopedia, 27 Sep. 2018. Web. 17 Oct. 2018.

[34] Thorlabs. *'Collimation Tutorial: Choosing a collimation lens for your laser diode'*, 2018. Thorlabs Inc.

[35] Newport. *'Optics Tutorial: Light Collection and Systems Throughput'*. 2018. Newport Corporation.

[36] Thorlabs. *'Lens Tutorial'*, 2018. Thorlabs Inc.

[37] R. Paschotta, article on 'diffraction-limited beams' in the *Encyclopedia of Laser Physics and Technology*, 1. edition October 2008, Wiley-VCH, ISBN 978-3-527-40828-3

[38] Richmond D Richard, Cain C Stephen. Direct Detection LADAR Systems. SPIE PRESS. 2010. Print

[39] Burns, Hoyt N., Christos G. Christodoulou, and Glenn D. Boreman. "System design of a pulsed laser rangefinder." *Optical Engineering* 30.3 (1991): 323-330.

[40] Newport. *'Introduction to Solar Radiation'*. 2018. Newport Corporation.

[41] R. M. Measures, *Laser Remote Sensing: Fundamentals and Applications*. (Wiley, New York, 1984).

[42] SensL (Sense Light). *'Ranging Demonstrator Description'*. March 2016

[43] Edmund Optics. *'Understanding Focal length and Field of view'* 2018 Edmund Optics Inc.

[44] Newport. *'Technical Note: Optics Fundamentals'*. 2018. Newport Corporation

[45] Peter Kaldén, Erik Sternå (Master's Thesis 2015) "Development of a low-cost laser range-finder (LIDAR)" *Department of Microtechnology & Nanoscience, Chalmers University of Technology, Gothenburg, Sweden.*

[46] Pedrotti, Frank L., Leno M. Pedrotti, and Leno S. Pedrotti. *'Introduction to optics'*. Cambridge University Press, 2017.

[47] Pacala, Angus, Tianyue Yu, and Louay Eldada. *"Robust lidar sensor for broad weather, shock and vibration conditions."* U.S. Patent Application No. 14/140,522

[48] Ready, J. (1999). Optical Detectors and Human Vision. *Fundamentals of photonics*, 211-248.

[49] Thorlabs. *'Photodiode-Tutorial'*. 2018.Thorlabs Inc.

[50] Grund, Christian J., et al. "High-resolution Doppler lidar for boundary layer and cloud research." *Journal of Atmospheric and Oceanic Technology* 18.3 (2001): 376-393

[51] Hardesty, R. Michael, W. Alan Brewer, and B. J. Rye. "*Estimation of wind velocity and backscatter signal intensity from Doppler lidar returns.*" Conference Record of the Thirty-First Asilomar Conference on Signals, Systems and Computers (Cat. No. 97CB36163). IEEE, 1997.

[52] Boyd, Robert W. "*Radiometry and the detection of optical radiation.*" *New York, John Wiley and Sons, 1983, 261 p.* (1983).

[53] Newport. *'User's Manual (818 Series Photodetector Guide)'* 2018. Newport Inc.

[54] Kaufmann, K. (2005). *'Choosing your detector'*. Hamamatsu Photonics.

[55] Wang, W. C. (2011). *'Optical Detectors'*. Seattle: Department of Mechanical Engineering-University of Washington.

[56] Williams, George M., Madison A. Compton, and Andrew S. Huntington. "*Single-photon-sensitive linear-mode APD Ladar receiver developments.*" *Laser Radar Technology and Applications XIII*. Vol. 6950. International Society for Optics and Photonics, 2008.

[57] Velodyne. *'Eye-Safety Resource Manual'*. April 2007. Velodyne Inc.

[58] Skolnik, M., "Introduction to radar", Radar handbook **2** (1962).

[59] Williams, G., "Optimization of eye safe avalanche photodiode lidar for automobile safety and autonomous navigation systems", Optical Engineering **56**, 3, 031224 (2017).

[60] Williams, G., "Range-Walk Correction Using Time Over Threshold", Tech. rep., Voxtel, Inc (2018).

[61] Ruotsalainen, T., P. Palojarvi and J. Kostamovaara, "A wide dynamic range receiver channel for a pulsed time-of-flight laser radar", IEEE Journal of Solid-State Circuits **36**, 8, 1228–1238 (2001).

[62] Kurtti, S. and J. Kostamovaara, "Pulse width time walk compensation method for a pulsed time-of-flight laser rangefinder", in "Instrumentation and Measurement Technology Conference, 2009. I2MTC'09. IEEE", pp. 1059–1062 (IEEE, 2009).

[63] Kurtti, S. and J. Kostamovaara, "An integrated laser radar receiver channel utilizing a time-domain walk error compensation scheme", IEEE Transactions on instrumen- tation and measurement **60**, 1, 146–157 (2011).

[64] Nissinen, J., I. Nissinen and J. Kostamovaara, "Integrated receiver including both receiver channel and TDC for a pulsed time-of-flight laser rangefinder with cm- level accuracy", IEEE journal of solid-state circuits **44**, 5, 1486–1497 (2009).

[65] Nakhostin, M., *Signal Processing for Radiation Detectors* (John Wiley & Sons, 2017).

[66] Neyman, J. and E. Pearson, "On the problem of the most efficient tests of statistical inference", Biometrika A **20**, 175–240 (1933).

[67] Kay, S., "Fundamentals of statistical signal processing, Volume I: Estimation theory (v. 1)", PTR Prentice-Hall, Englewood Cliffs (1993).

[68] Kay, S., *Fundamentals of Statistical Signal Processing: Detection theory*, Prentice Hall Signal Processing Series (Prentice-Hall PTR, 1998).

[69] Cronologic, "Ndigo5G-8, Ndigo5G-10 User Guide", (2018).

[70] Khader, Motaz and Samer Charlan. *An introduction to automotive LIDAR*. Texas Instruments, October 2018.

[71] Rablau, Corneliu. "Lidar: a new self-driving vehicle for introducing optics to broader engineering and non-engineering audiences," *Proc. SPIE 11143, Fifteenth Conference on Education and Training in Optics and Photonics: ETOP 2019.*

[72] Mirrorcle. *'Mirrorcle Technologies MEMS Mirrors – Technical Overview'*. 2019. Mirrorcle Technologies.

[73] Mirrocle. *'USB-SL MZ USB MEMS Controller User Guide'*. 2019. Mirrocle Technologies.

[74] Royo, Santiago and Maria Ballesta-Garcia. "An Overview of Lidar Imaging Systems for Autonomous Vehicles." *Appl. Sci.* 9, no. 19: 4093.

www.ingramcontent.com/pod-product-compliance
Lightning Source LLC
Chambersburg PA
CBHW020545220526
45463CB00006B/2196